T0240051

Artificial Intelligence in Daily Life

Raymond S. T. Lee

Artificial Intelligence in Daily Life

Raymond S. T. Lee
Division of Science and Technology
Beijing Normal University-Hong Kong Baptist University United
International College
Zhuhai, Guangdong, China

ISBN 978-981-15-7697-3   ISBN 978-981-15-7695-9   (eBook)
https://doi.org/10.1007/978-981-15-7695-9

This Springer imprint is published by the registered company Springer Nature Singapore Pte Ltd.
The registered company address is: 152 Beach Road, #21-01/04 Gateway East, Singapore 189721, Singapore

This book is dedicated to all readers and students taking my courses in AI and Data Mining, your enthusiasm in learning new concepts and seeking knowledge prompted me to write this book.

Preface

Motivation for This Book

Artificial Intelligence (AI) technology and its related applications become part of life in ways that we could not even think of a century ago. With the exponential growth of AI in the past decades, our routines have changed beyond measures due to robotics and AI that are used in a wide array of day-to-day services. Despite AI being at its infancy, we have already been benefited immensely.

Contemporary books and research monographs on AI are either (i) *technical books* written by academics from the fields of Computer Science and Information Technology which mainly focus on AI mathematical models and derivations. As a result, they are usually too difficult and mathematically complex for readers of different educational backgrounds to comprehend; or (ii) *popular science type of books* written by authors from other non-computer science fields which describe AI as high-level concepts and topics such as AI philosophy and ethics, AI and conscious machine, lack of basic concepts' overview, AI core technology, and how AI applications reshape our daily life.

The motivation and main purpose of this book is to provide AI basic concepts and knowledge for both Computer Science (CS), non-CS students and readers to understand how AI technology such as *software robots, natural language processing (NLP), semantic and ontological-based search engine, intelligent city*, and *intelligent campus* are applied to daily life and activities.

Organization of This Book

For the ease of readership, this book consists of four different parts. They include

Part I AI Concepts
   Discusses AI basic concepts and history.
Part II AI Technology
   Discusses five core AI technologies which are the building blocks for different kinds of AI applications: *machine learning (ML), data mining (DM), computer vision (CV), natural languages processing (NLP), ontological-based search engine (OSE).*
Part III AI Applications
   Discusses major contemporary AI applications that affect ways of living, working style, and environment ranging from *intelligent agents* and *software robots* to *intelligent transportation systems (ITS)* and *smart cities.*
Part IV Beyond AI
   Discusses topics beyond but critically important AI future development including AI ethics, conscious mind development, autonomous robotics in daily activities, and related topics.

This book is organized as follows:

- Chapter 1—A Brief Journey of Human Intelligence
  Human intelligence is a cross-discipline study that puzzles philosophers, natural scientists, psychologists, cognitive and neuroscientists for centuries. As the introductory chapter of the book, this chapter explores *human intelligence* in various aspects including Greek mythology, philosophy, psychology, cognitive science, and neuroscience.
- Chapter 2—AI Fundamentals
  This chapter introduces the main theme of this book—Artificial Intelligence (AI). We begin with a general definition and its interpretation followed by a brief history. Next, we discuss one of the most interesting experiments in AI—the *Turing Test*. Then, we discuss two of AI's main themes: *Strong AI versus Weak AI*. After that, we introduce five of AI's main components: *machine learning (ML), data mining (DM), computer vision (CV), natural language processing (NLP), and ontological-based search engine (OSE).*

- Chapter 3—Machine Learning (ML)
  This chapter explores the first and foremost AI component and technology—*machine learning (ML)*. We explore how humans learn and discuss three major machine learning models: *supervised learning, unsupervised learning, and reinforcement learning*. We also study how the human brain works to *think and learn—biological neural networks* which leads to the design of *artificial neural networks (ANN)*—mathematical and computational counterparts to simulate human memory, learning, and thinking processes.
- Chapter 4—Data Mining (DM)
  This chapter explores various methods and technologies involving *data mining* that includes *KNN* for clustering, *decision tree* for decision-making, *regression* for forecast and projection, and *association rule* for mining useful patterns. We also introduce *deep neural networks (DNN)* on data mining and how these technologies can be applied to various real-world problems.
- Chapter 5—Computer Vision (CV)
  This chapter compares the *human visual system* with *computer vision*. By imitating human vision, we introduce 3 computer vision components: (1) *Figure-scene segmentation*, (2) *Object recognition*, and (3) *3D and VR modelling*. After that, we study various latest computer vision technologies and applications that are related to daily life and activities.
- Chapter 6—Natural Language Processing (NLP)
  This chapter begins with the introduction of *human language* and *intelligence*. We will also introduce the *six linguistics levels* in human languages. Next, we study NLP main components including *natural language understanding (NLU), speech recognition, syntactic analysis, semantic analysis, pragmatic analysis*, and *speech synthesis* followed by major NLP applications related to our daily life.
- Chapter 7—Ontological-Based Search Engine (OSE)
  The first part of this chapter begins with the main components of *traditional search engine*, its major shortcomings in terms of system and user perspectives, and study on several commonly used search engines' basic architectures. The second part of this chapter introduces *ontological-based search engine (OSE)*. It begins with *knowledge* and *ontology* basic concepts followed by *ontology engineering, semantic web*, and how to use *ontology graph (OG)* to represent concepts and ideas. Then, we will study OSE system architecture with its several major applications including *intelligent content management system, news retrieval*, and *ontological-based search engine* and *web ontology learning systems*.

- Chapter 8—Intelligent Agents and Software Robots

  This chapter introduces *intelligent agents (software robots)*, their basic requirements, and explores different varieties of intelligent agent frameworks followed by several major applications in daily activities including *agent shoppers, agent negotiator, agent weatherman*, and *agent traders*. Lastly, we explore the threats and challenges of intelligent agent technology.

- Chapter 9—Intelligent Transportation

  This chapter begins with a general transportation system overview. Next, we study the major component of *intelligent transportation—5G technology* and a review from *1G to 5G* in the past half century. Next, we study *intelligent transportation system (ITS)* and its potential applications. Then, we examine 5G-enabled ITS technology—*V2X (Vehicle-to-Everything)* with related technologies such as *V2V (Vehicle-to-Vehicle)* and *V2I (Vehicle-to-Infrastructure) Technologies*. After that, we study 5G and AI technology integration for the implementation of new age ITS applications including *smart cities, autonomous vehicles, intelligent traffic management systems, emergency services*, and *future-proof infrastructure* that reshape our daily activities.

- Chapter 10—Smart Health

  This chapter gives an *IoT (Internet of Things)* overview and *wearable computing technology* from the blueprint of *Smart Health*. We will also learn the latest *wearable healthcare device* invention ranging from a *smart watch* for heart rate and blood pressure to *biosensor* healthcare monitoring services. After that, we will study two innovative AI technologies applied to healthcare: (1) *health chatbot* and (2) *robot-assisted surgery (RAS) technology*.

- Chapter 11—Smart Education

  This chapter begins with *smart education* progress in the past decades. Next, we examine *smart education model*—a four-tier framework of *smart pedagogies* and key features of *smart learning* environment. Then, we study two latest R&D AI-based smart education applications: *AI language learning robots* and *VR-AR teacher* to elaborate on how AI technology we learnt in the previous chapters such as machine learning, NLP technology, ontological knowledgebase, VR and AR can be integrated to provide a new age of *smart education*.

- Chapter 12—Smart City

  This chapter begins with *smart city* definition and why we need it. Next, we explore its major components and infrastructure. Next, we learn different countries' progress on *smart city* in the past decades. Then, we examine four critical masses of smart city: *smart transportation, smart energy, smart health care*, and *smart technology*. After that, we study three major supporting technologies *IoT, big data, and AI* with several innovative applications/systems including *smart pole, smart house*, and *smart campus*.

- Chapter 13—AI and Self-consciousness
  This chapter explores AI ultimate question and challenge—*self-consciousness and self-awareness*, how intelligent robots (or AI systems) can have so-called *subjective experiences*. We begin with *consciousness concepts* and *machine consciousness* in neuroscience disciplines' brief literature review to current AI and machine learning R&D. Next, we explore *machine consciousness* typical approach—the *Good Old-Fashioned Artificial Consciousness (GOFAC)* which consists of five major components: (1) *Functionalism*; (2) *Information integration*; (3) *Embodiment*; (4) *Enaction*; and (5) *Cognitive mechanisms*. Lastly, we will conclude AI and machine consciousness study, outstanding issues, and problems to approach in order to design and build a truly self-consciousness and self-awareness robot.

- Chapter 14—AI Ethics, Security, and Privacy
  This chapter explores one of AI and robotics' popular and controversial topic—*AI ethics*. First, we introduce AI ethics with *Asimov's Three Laws of Robotics*. Next, we study major aspects and concerns related to AI ethics including robot ethics, robot rights, moral agents, opaqueness of AI systems, privacy and AI monitoring, automation and employment, prejudices in AI systems, responsibility for autonomous machines, and international AI ethic policy.

- Chapter 15—What's Next?
  As the closing chapter of the book, this chapter explores several AI cutting-edge-related technologies: *singularity* and *superintelligence, quantum computing*, and *6G technologies*. First, we begin with singularity in *AI and superintelligence* concepts and major concerns. Next, we present *quantum computing* basic concepts and how such innovative theories are truly related to AI technology. Then, we explore technology beyond 5G—*the 6th Generation Communication Technology*, its major features, and how these three powerful technologies can be integrated to form a new AI and *smart city* era. Lastly, we conclude the book with closing remarks—*the future is our choice*.

Readers of This Book

This book is both a textbook and a general IT-related book tailored for

- Undergraduate students for various courses and disciplines to learn AI basic concepts, technologies, and applications.

- Lectures and tutors who would like to teach and organize courses with tutorials to teach undergraduate students on AI general knowledge and introduction, with the potential and AI applications that are related to daily activities.
- Readers of different backgrounds and disciplines (non-mathematical- and non-CS-based) to learn AI knowledge and basic concepts, the core technology, and how AI can be used in daily activities.

How to Use This Book?

This book can be served as a textbook for undergraduates of various disciplines and as a general IT-related book for readers to learn AI key concepts, technologies, and applications.

Part I (Chaps. 1–2) covers AI basic concepts and brief history. Part II (Chaps. 3–7) covers five major AI technologies: machine learning, data mining, computer vision, natural language processing, and ontological-based search engine. Part III (Chaps. 8–12) covers major AI applications in our daily activities. Part IV (Chaps. 13–15) covers topics beyond but critically important for AI future development including AI and self-consciousness, AI ethics, and super intelligence.

In UIC, this book is served as the core textbook tailored for General Education (GE) undergraduate course *AI in Daily Life* for all undergraduate students of the university. The book is designed for a 14-week 2-hour lecture and 1-hour tutorial on a case study.

This book can be also served as year 1 *Foundation Course of AI* for undergraduate/postgraduate students of computer science, AI, and data science programmes.

For non-AI major research students and data scientists, this book can be served as an introduction to learn basic concepts, components, and core applications related to AI, robotics, smart cities, and smart applications.

For general readers of various backgrounds who would like to learn AI basics and related technologies, this book can be regarded as an introductory and background reading of AI and related technologies.

Zhuhai, China   Raymond S. T. Lee

Acknowledgements

While it took me around 6 months to write this book, my whole journey of AI started almost 30 years ago when I was still an undergraduate at University of Hong Kong (HKU) studying AI and networking courses since 1986.

I would like to express my gratitude to the following people for their support and assistance.

To my wife Iris for her patience, encouragement, and understanding, especially during my time spent on research and writing in the past 30 years.

To Ms. Celine Cheng, Executive Editor of Springer Nature and her professional editorial and book production team members Ms. Jane Li and Mr. Karthik Raj for their support, valuable comments, and advice.

To Prof. James Liu, my M.Sc. and Ph.D. mentor, for his support and for leading me to the *sacred land* of AI.

To team members, research assistants, and research students of UIC iCampus R&D projects of ITSC and Innovative Center of UIC for their support and assistance.

To Prof. Tang Tao, President of UIC, for the provision of an excellent environment for the research, teaching, and the writing of this book.

To Prof. Zhi Chen, Provost of UIC and Prof. Weijia Jia, Vice President (Research and Development) of UIC for their support for the R&D of iCampus projects in UIC.

To Prof. Hua Xiong Huang, Vice President (Academic) and Dean of Division of Science and Technology of UIC, and Prof. Weifeng Su, Programme

Director of Computer Science and Technology of UIC, for their support for the opening of the new course *Artificial Intelligence in Daily Life* in UIC.

To UIC, for the prominent support on research grant #R202008, and for the provision of an excellent environment and computer facilities for the preparation of this book.

Zhuhai, China   Raymond S. T. Lee
June 2020

About This Book

This book consists of four parts. Part I—AI Concepts, which discusses basic concepts and the history of AI. Part II—AI Technology, which discusses the core AI technologies. Part III—AI Applications, which discusses contemporary and major AI applications that affect daily activities. Part IV—Beyond AI, which discusses topics beyond but critically important for AI future development.

The aims of this course are

(1) To teach students and readers AI basic concepts and knowledge.
(2) To teach students and readers the core AI technology and how they are used in daily activities.
(3) To teach students and readers how AI technology can be used for their academic studies and future works.
(4) To teach students and readers major AI applications that reshape present and future daily activities.
(5) To teach students and readers basic AI applications using case studies on different contemporary AI-related projects and applications such as intelligent agents, NLP-based software robots, semantic and ontological-based search engines, intelligent campus and intelligent city, and how they affect daily activities.

This book is both a textbook and a general IT-related book tailored for

(1) Undergraduate students for various courses and disciplines to learn AI basic concepts, technologies, and applications.
(2) Lectures and tutors who would like to teach and organize courses with tutorials to teach undergraduate students on AI general knowledge and introduction, with the potential and AI applications that are related to daily activities.
(3) Readers of different backgrounds and disciplines (non-mathematical- and non-CS based) to learn AI basic knowledge and concepts, the core technology, and more importantly, how AI can be used in daily activities.

Contents

Part IV   Beyond AI

About the Author

Raymond S. T. Lee founder of quantum finance forecast center with over 20 years of IT consultancy, R&D experiences in AI, chaotic neural networks, intelligent fintech system, quantum finance, and intelligent e-commerce systems had successfully commercialized his AI-Fintech invention in business sectors in China and Hong Kong. Dr. Lee attained his B.Sc. (Physics) from Hong Kong University in 1989, M.Sc. (Information Technology), and Ph.D. (Computer Science) from Hong Kong Polytechnic University in 1997 and 2000, respectively. After graduating from Hong Kong University, he joined the Hong Kong Observatory of the Government of the Hong Kong Special Administrative Region as a meteorological scientist on weather forecasting and developing a numerical weather forecast system from 1989 to 1993.

From an academic perspective, Dr. Lee had worked at the Department of Computing of Hong Kong Polytechnic University (HKPolyU) as a Lecturer, was promoted as Assistant Professor in 2000 and Associate Professor in 2005, respectively. During this time, he had published over 90+ publications and authored six textbooks and monographs covering the fields at AI, chaotic neural networks, AI-based fintech systems, intelligent agent technology, chaotic cryptosystems, ontological agents, neural oscillators, biometrics, weather simulation and forecasting systems.

From a commercial perspective, Dr. Lee was invited to join Leanda Investment Group in China (2012–2017) as Group CTO/Chief Analyst to implement his AI-Fintech invention—Quantum Finance Forecast System on major commodities in China for 1000+ investors. In March 2017, he set up the

Quantum Finance Forecast Center (QFFC) (http://qffc.org), a nonprofit-making, AI-Fintech R&D and worldwide financial forecast center aiming at the R&D and provision of a free and open platform for worldwide traders and individual investors to acquire free knowledge of worldwide 129 financial product forecasts based on the state-of-the-art AI, chaotic neural networks, and quantum field theory technologies.

Upon the completion of the QFFC project on the automation of the Quantum Finance Forecast System, Dr. Lee joined United International College (UIC) in China to further his R&D works on AI-Fintech and contribute his knowledge on AI-Fintech, chaotic neural networks, and related intelligent systems to fellow students and the community. His latest book Quantum Finance: Intelligent Financial Forecast and Quantum Trading Systems which serves as a textbook for the new course Quantum Finance in UIC was published by Springer Nature by in Jan 2020.

Part I

AI Concepts

1

A Brief Journey of Human Intelligence

One day, however, I heard someone reading from a book he said was by Anaxag-oras, according to which it is, in fact, intelligence that orders and is the reason for everything. Now this was a reason that pleased me; it seemed to me, somehow, to be a good thing that intelligence should be the reason for everything. And I thought that, if that's the case, then intelligence in ordering all things must order them and place each individual thing in the best way possible; so if anyone wanted to find out the reason why each thing comes to be and perishes or exists, this is what he must find out about it: how is the best for that thing to exist, or to act or be acted upon in any way? On this theory, then, a person should consider nothing else, whether in regard to himself or anything else, but the best, the highest good; though the same person must also know the worst, as they are ob-jects of the same knowledge. Reckoning thus, I was pleased to think I'd found, in Anaxagoras, an instructor in the reason for things to suit my own intelligence.
Phaedo (97c-d,) Plato
(Plato et al. 1998)

Abstract *Human intelligence* is a cross-discipline study that puzzles philosophers, natural scientists, psychologists, cognitive scientists, and neuroscientists for centuries. As the introductory chapter of the book, this chapter gives an overview on the source of AI—human intelligence, a foundation research topic developed by distinguished philosophers including Plato, Aristotle, Descartes, Kant, etc. It presents a thorough, cross-discipline exploration ranging from Greek mythology, philosophical schools of thought regarding human knowledge and intelligence to contemporary theories and studies on human intelligence in the areas of psychology, cognitive science, and neuroscience.

1.1   What is Intelligence?

Artificial intelligence (AI) has been studied for centuries. It not only involves the field of computer science, but is also closely related to other disciplines, such as philosophy, psychology, epistemology, neuroscience, and neurophysiology.

It is natural and critical to understanding the fundamental issues about intelligence before we explore AI territory (Lee 2006):

- What is intelligence?
- What are the main schools of thought and approaches to study intelligence?
- How can we interpret and measure intelligence?
- How does our brain work to facilitate intelligence?

This chapter gives an overview of the source of AI—human intelligence, a foundation research topic developed by distinguished philosophers including Plato, Aristotle, Descartes, Kant, etc. It presents a thorough, cross-discipline exploration ranging from Greek mythology, philosophical schools of thought regarding human knowledge and intelligence to contemporary theories and studies on human intelligence in the areas of psychology, cognitive science, and neuroscience (Carter 2007; Sternberg and Kaufman 2011).

1.2   Greek Mythology—Prometheus

Many textbooks named the origin and birth of AI either from Dartmouth's meeting or the famous *Turing Test* invented by Sir Alan Turing in 1950.

Is it? The answer is *yes* and *no*.

Yes in the sense that five AI founders including Prof. Allen Newell, CMU (1927–1992), Prof. Herbert Simon, CMU (1916–2001), Prof. John McCarthy, MIT (1927–2011), Prof. Marvin Minsky, MIT (1927–2016), and Prof. Arthur Samuel, IBM (1901–1990) coined the word *Artificial Intelligence (AI)* at this first meeting in 1956, or Sir Alan Turing invented the famous *Turing Test* as an AI formal test in 1950 (Russell and Norvig 2016).

No in the sense that the notion of *man-made intelligence* or what we called AI is not a new concept.

In Greek mythology, epic poet Hesiod's Theogony in 700 B.C. narrated *Prometheus*, a Titan and culture hero who created the first human from clay (a *humanoid robot* in today's term).

Prometheus shaped human after God's image and allowed it to walk upright that might look toward the heavens (Buxton 2004). He determined to *upgrade* human minds and their conditions by defying Zeus' will to steal *fire (intelligence)* from the chariot of the sun, a gift unknown to mankind.

Fire bestowed by Prometheus marked the beginning of civilization. He taught them to survive in harsh weather conditions, craft tools for agriculture, build weapons to defend against wild animals, and other skills; eventually, human became superior at any rate and thrived (Fig. 1.1).

1.3 Definition of Human Intelligence

One of the widely accepted definitions of human intelligence in the past century is a statement condensed by Emeritus Prof. Linda Gottfredson (1997) published at the editorial of *Journal Intelligence* in 1997 with endorsements from 52 distinguished AI researchers and professors (Fig. 1.2).

Intelligence is a very general mental capability that, among other things, involves the ability to reason, plan, solve problems, think abstractly, comprehend complex ideas, learn quickly and learn from experience. It is not merely book learning, a narrow academic skill, or test taking smarts. Rather, it reflects a broader and deeper capability for comprehending our surroundings – "catching on", "making sense" of things, or "figuring out" what to do …
Gottfredson, 1997

1.4 Philosophical View of Human Intelligence—Mind–Body Dualism

The study of human intelligence had a long history dating back to the Ancient Greeks. Plato (428–347 B.C.) was the earliest Western philosopher who considered the idea of *Mind–Body Dualism* that human intelligence as the faculty of *mind* or *soul* should be independent of our physical body (Bussell 2010).

Fig. 1.1 Prometheus brings fire (intelligence) to humankind (Wikimedia Commons 2020a) (Public Domain Mark 1.0)

Philosopher, mathematician, and scientist René Descartes (1596–1650) further extended such idea and described the *mind* as a non-extended, nonphysical substance, the so-called *res cogitans*.

His works *Discourse on Method and Meditations on First Philosophy* (1641) described that the mind possesses *consciousness* and *self-awareness*, to distinguish it from the brain as what we call *intelligence* nowadays. One of his famous quotations in *cogito ergo sum* (English: *I think, therefore I am*) concluded the fact that: *If he doubted, then something or someone must be doing the doubting. Therefore, the very fact that he doubted proved his existence.*

Fig. 1.2 Seeking for human intelligence (Tuchong 2020a)

He is in fact the first philosopher who formulated the *mind–body problem* in the form is still in use today (Descartes and Clarke 2003) (Figs. 1.3 and 1.4).

1.5 Philosophical View of Human Intelligence—Kant and Priori Knowledge

There is one mystery troubling philosophers for many years is the problem of *priori knowledge* (A priori in philosophical terms), considered by one of the most influential philosophers in history. Immanuel Kant (1724–1804) in his famous work *Critique of Pure Reason* published in 1781 (Kant 1998) pointed out that the word *critique* not only means denying all events, but also means critical thinking and reasoning, whose results can be positive or negative. It is purely a technical term defined and does not contain any content derived from our experience. This is an important and basic concept in his prior knowledge exploration. *Reason* is another technical term defined by him. It is a conceptual element in cognition. We introduce it into *experience* rather than derive it from it—it represents a *priori concepts*. In other words, a *priori*

Fig. 1.3 Plato (428–347 B.C.) (Tuchong 2020b)

knowledge is independent of experience and is inferred from what we call *pure reason*. The posterior knowledge depends on the experience or empirical evidence of most natural science and *personal knowledge* aspects. One of his main contributions is his unique interpretation of a *priori knowledge*. He believes that we have the ability to obtain a *priori knowledge* and determine the authenticity of such knowledge. He believed firmly in our mental processes such as *thinking* and *perception,* so we automatically manipulate all basic elements, including space, time, logic and matter, and causality to achieve our goals (Fig. 1.5).

1.6 Psychological View of Human Intelligence

Modern psychology believes that *human intelligence* should be evaluated and determined in all aspects of mental ability and ability that includes verbal and perception abilities, memory ability, and manipulation speed. The *Webster Adult Intelligence Scale (WAIS)* is an IQ test designed to measure intelligence

Fig. 1.4 René Descartes (1596–1650) (Tuchong 2020c)

and cognitive abilities of adults and adolescents. The original *WAIS (Form I)* was published by Prof. David Wechsler (1896–1981) in February 1955. It is a revised version of the *Wechsler–Bellevue Intelligence Scale* released in 1939. The fourth version *WAIS-IV* released by Pearson in 2008 is used worldwide as the most extensive IQ test (Kaufman and Lichtenberger 2006). The current version of *Webster's Adult Intelligence Scale IV Test* assesses general human intelligence (the so-called *g-value*) by

- Verbal comprehension index scale,
- Perceptual reasoning index scale,
- Working memory index scale, and
- Processing speed index scale.

The current version of the test, WAIS-IV released in 2008, composed of 10 core subtests and 5 supplemental subtests, with the 10 core subtests yielding scaled scores that sum to derive *Full-Scale IQ*. The *General Ability Index (GAI)* was incorporated with *similarities, vocabulary,* and *information*

Fig. 1.5 Immanuel Kant (1724–1804) (Wikimedia Commons 2020b) (Public Domain Mark 1.0)

subtests from *Verbal Comprehension Index* and *block design, matrix reasoning, and visual puzzles subtests* from the *Perceptual Reasoning Index*. Figure 1.6 shows a WASI-IV Chart.

1.7   Cognitive Scientific View of Human Intelligence

Cognitive science is an interdisciplinary scientific study of psychology and its processes. It broadly examines the nature, tasks, and *cognition* functions. The intelligence and behavior studied by cognitive scientists focus on how the nervous system represents, processes, and transforms information. Cognitive science is one of the earliest disciplines that provides scientific methods to understand, interpret, and evaluate human intelligence. It involves philosophy, psychology, artificial intelligence, neuroscience, linguistics, and anthropology; most of which are core components and research areas related to AI

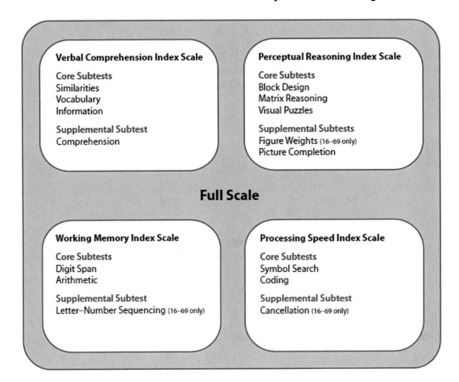

Fig. 1.6 WASI-IV chart for general human intelligence measurement

research today. Its ideas originated when researchers in various fields began to develop the *Theory of Mind* based on complex representations and calculation procedures in the mid-1950s. The *MIT Encyclopedia of Cognitive Sciences* (Wilson and Keil 1999) defines intelligence as *the ability to adapt, shape, and choose the environment*. In cognitive science, artificial intelligence involves the study of *cognitive phenomena in machines*. One of the practical goals is to realize human intelligence in computers. Computers are also widely used as tools for studying cognitive phenomena. Figure 1.7 illustrates areas that contribute to cognitive science development (Bermúdez 2020).

In recent centuries, many schools of thought have interpreted and studied human intelligence from the perspective of cognitive science including

- The psychological method proposed by Sir Francis Galton (1822–1911) in 1883 based on typical psychological skills, for example, *Just Noticeable Difference (JND)* method to evaluate human intelligence (Galton 2012).

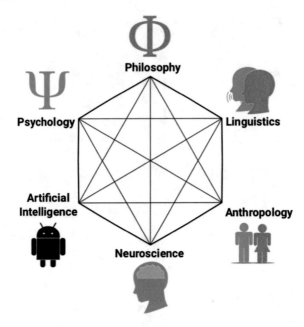

Fig. 1.7 Major fields contributing to cognitive science development

- Professor Charles Spearman (1863–1945) in his extraordinary book *Essence of Intelligence and Cognitive Principles* (Spearman 1923) interpreted intelligence into three main psychological activities: understanding of experience, education of correlates and relations.
- The founder of developmental psychology Prof. Jean Piaget (1896–1980) in his famous book *Psychology of Intelligence* (Piaget 1950) proposed a variety of intellectual viewpoints including biological adaptability and the famous *Gestalt Psychology* and human intelligence theory.
- Professor Alfred Binet (1857–1911) and Prof. Théodore Simon (1873–1961) (1916) interpreted intelligence as *Complex Judgment Ability* (CJA) composed of 3 core cognitive abilities: directional, adaptable, and regulation abilities.
- The latest research on cognitive methods of intelligence include Prof. Robert Sternberg's *Cognitive Composition Method* (Sternberg, 1977) in his outstanding book *Intelligence, Information Processing, and Analogical Reasoning: The Componential Analysis of Human Ability* (Sternberg 1977) interpreted human intelligence as information processing components with basic but complex reasoning and problem-solving tasks such as analogy, syllogism, verbal understanding, nonverbal cue decoding, and future-event prediction.

- Professor Howard Gardner, a remarkable cognitive psychologist of twentieth century put forward an innovative perspective on human intelligence understanding, namely *Multiple Intelligence (MI)*. In his extraordinary book *Frame of Mind: The Theory of Multiple Intelligence* (Gardner 2011), he questioned human intelligence unity and suggested to interpret intelligence as the integration of eight different aspects: logical–mathematical intelligence, linguistic intelligence, spatial intelligence, musical intelligence, bodily–kinesthetic intelligence, interpersonal intelligence, naturalistic intelligence, and intrapersonal intelligence. More importantly, he believes that each of these intelligent components is independent of the other components in some way.

1.8 Neuroscience View of Human Intelligence

Neuroscience is the scientific study of the *nervous system*. Neuroscientists focus on the brain and its influence on behavior, cognitive functions, and people's way of thinking. The latest neuroscience tells us that our brain is composed of more than 10^{11} neurons. The way these neurons are organized is a complex problem itself, not to mention the study of how these neurons work together to perform the *thinking* and *learning processes*.

The first scientific work in the field of *brain science* (the structure and nerve function of the brain) was by Prof. Camillo Golgi (1843–1926) who invented the *staining method* to study brain neural activity. He accidentally discovered that only some of the brain cells turned black by using silver salt to stain meninges, while most of them remained unstained. Based on this major discovery, he proposed that the brain is composed of a sponge-like structure called *syncytia*. As we now know, these stained tissues are *neuronal cells (neurons)* activated by staining rather than by sponge-like tissues (Gardner 1993). Nonetheless, his discovery provided a key breakthrough in understanding the neural structure of the brain. Neuroanatomist Prof. Santiago Ramón y Cajal (1852–1934) proposed an innovative idea based on Golgi's staining method. He concluded that: *these stained tissues are not sponge-like elements, but a collection of brain cells called neurons, which are connected together to form a complex network structure—the so-called biological neural network as we all know nowadays* (Fig. 1.8).

Figure 1.9 illustrates a simplified diagram of a *biological neural network* (Lee 2006). Each nerve cell (neuron) consists of

Fig. 1.8 Biological neural networks (Tuchong 2020d)

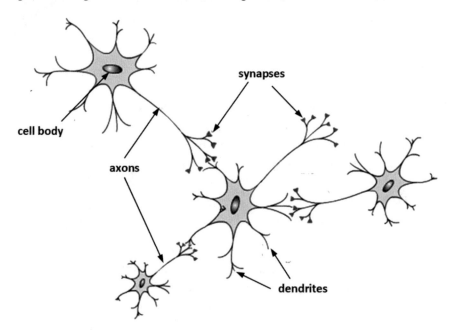

Fig. 1.9 Simple illustration of biological neural networks

- *Axon,* an elongated filament that broadly branches from its neuron to connect to another neuron.
- *Dendrites,* a dendritic structure branching from a neuron, intercepting the stimulation of other neurons like a sensor.
- *Nucleus,* the central body of neurons, embedded in cytoplasm.
- *Synapse,* the *axon tip* that connects other neurons by attaching to neighboring neuron dendrites.

In fact, each neuron has more than 1,000 synapses on the dendritic tree, and neurophysiologists today pointed out that the major function for these dendritic structures is to maximize the number of contact points between a neuron and its neighboring neurons in order to facilitate effective *information processing* and *transmission.*

How does this neural network play a role in our *thinking process?*

Neurophysiologist Prof. Warren McCulloch (1898–1969) and mathematician Prof. Walter Pitts (1923–1969) published an influential paper on neural activity in 1943, *Bulletin of Mathematical Biophysics: A logical calculus of the ideas immanent in nervous activity* (McCulloch and Pitts 1943). They suggested that the main function of neural activity is to *process information,* and not to use *Energy Theory* to explain the neural activity. They thought that neurons function like *logical switches* in electronic devices. They also illustrated how to use their proposed neural network to perform basic logical operations, such as *AND, OR, and NOT* in this influential paper. Their contribution not only provided breakthroughs in neurophysiology, but also a solid foundation for the development of *digital computers.* Although we now know that neurons' characteristics are different from logical switches, neurons are like nonlinear *integrate-and-fire operations* that transfer and store information (Freeman 2002).

1.9 Conclusion

Before we begin the journey of Artificial Intelligence (AI), we learnt human intelligence from various aspects covering Greek mythology, philosophy, psychology, cognitive science, and neuroscience. Human intelligence is a cross-discipline study that puzzles philosophers, natural scientists, psychologists, cognitive scientists, and neuroscientists for centuries. The study of intelligence, primarily for human intelligence study and research, is a major focus not only on the mechanism of *how we think,* but is also based on human brain understanding, its thinking/memory manipulation and processing operations in order to shape our AI models. Furthermore, it also focuses on how to deploy these different models to construct AI systems and applications that can be used in our daily activities.

Plato believed that *intelligence* is the supreme thing, the thing to provide the reason for *everything*, and the belief that can show us the best way to understand everything we see and that occurs in the world. The author would also like to add that *intelligence is a gift to us. It is precious, it has the ability to lead us to the truth and reality. It is our free will to open our mind to see the truth* (Fig. 1.10).

Some people may think: *How far are we truly to understand intelligence?* The answer is that we wouldn't know, because the more we know about intelligence, the more we know many things that we don't really know. However, the author always believes that an *open mind to accept new concepts and ideas* is the key to opening the *door of intelligence*!

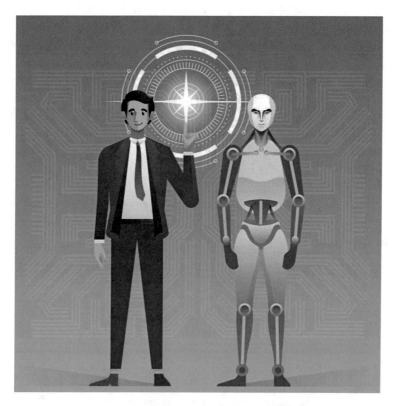

Fig. 1.10   From human intelligence to AI (Tuchong 2020e)

References

Bermúdez, J. L. (2020) *Cognitive science: An introduction to the science of the mind* (3rd ed.). Cambridge University Press.

Binet, A., & Simon, T. (1916). *The development of intelligence in children.* Baltimore: Williams and Wilkins.

Bussell, F. W. (2010). *Plato and the inherent dualism of scientific knowledge.* Kessinger Publishing.

Buxton, R. (2004). *The complete world of Greek mythology.* Thames & Hudson.

Carter, M. (2007). *Minds and computers: An introduction to the philosophy of artificial intelligence: An introduction to the philosophy of artificial intelligence.* Edinburgh: Edinburgh University Press.

Descartes, R., & Clarke, D. M. (2003). *Meditations and other metaphysical writings* (Penguin classics). Penguin Press.

Freeman, W. J. (2002). *How brains make up their minds.* New York: Columbia University Press.

Galton, F. (2012). *Inquiries into human faculty and its development.* Amazon.com.

Gardner, D. (1993). *Neurobiology of neural networks* (Computational Neuroscience). A Bradford Book.

Gardner H. (2011) *Frame of mind: The theory of multiple intelligence.* Basic Books.

Gottfredson, L. (1997). Mainstream science on intelligence: An editorial with 52 signatories, history, and bibliography. *Intelligence, 24*(1), 13–23.

Kant, I. (1998). *Critique of pure reason* (The Cambridge Edition of the Works of Immanuel Kant) (P. Guyer & A. Wood, Trans.). Cambridge University Press.

Kaufman, A. S., & Lichtenberger, E. O. (2006). *Assessing adolescent and adult intelligence* (3rd ed.). Hoboken (NJ): Wiley.

Lee, R. S. T. (2006). *Fuzzy-neuro approach to agent applications: From the AI perspective to modern ontology.* New York; Berlin: Springer.

McCulloch, W. S., & Pitts, W. (1943). A logical calculus of the ideas immanent in nervous activity. *Bulletin of Mathematical Biophysics, 5,* 115–133.

Piaget J. (1950) *The psychology of intelligence.* Taylor & Francis

Plato, et al. (1998). *Plato's phaedo.* Newburyport, MA: Focus Publishing.

Russell, S. J., & Norvig, P. (2016). *Artificial intelligence: A modern approach* (3rd ed.). Upper Saddle River, N.J; Harlow: Pearson Education.

Spearman, C. (1923). *The nature of intelligence and the principles of cognition.* Macmillan.

Sternberg, R. J. (1977). *Intelligence, information processing, and analogical reasoning: The componential analysis of human abilities.* Erlbaum

Sternberg, R. J., & Kaufman, S. B. (2011). *The Cambridge handbook of intelligence.* Cambridge, New York: Cambridge University Press.

Tuchong. (2020a). *Seek for human intelligence.* Retrieved May 3, 2020, from https://stock.tuchong.com/image?imageId=286594758192922638.

Tuchong. (2020b). *Plato portrait*. Retrieved June 5, 2020, from https://stock.tuc
hong.com/image?imageId=903207836876800018.

Tuchong. (2020c). *René Descartes portrait*. Retrieved June 5, 2020, from https://
stock.tuchong.com/image?imageId=79054181679422537.

Tuchong. (2020d). *Biological neural networks*. Retrieved May 3, 2020, from https://
stock.tuchong.com/image?imageId=483388184041619483.

Tuchong. (2020e). *From human intelligence to AI*. Retrieved May 3, 2020, from
https://stock.tuchong.com/image?imageId=486220165217517622.

Wikimedia Commons. (2020a). *Prometheus brings fire to mankind*. Retrieved May 3,
2020, from https://commons.wikimedia.org/wiki/File:Heinrich_fueger_1817_p
rometheus_brings_fire_to_mankind.jpg.

Wikimedia Commons. (2020b). *Immanuel Kant portrait*. Retrieved June 5, 2020,
from https://commons.wikimedia.org/wiki/File:Kant_gemaelde_3.jpg.

Wilson, R. A., & Keil, F. C. (Eds.). (1999). *The MIT encyclopedia of the cognitive
science*. MIT Press.

2

AI Fundamentals

Artificial intelligence, in fact, is obviously an intelligence transmitted by con-scious subjects, an intelligence placed in equipment. It has a clear origin, in fact, in the intelligence of the human creators of such equipment.
Pope Benedict XVI (Born 1927)

Abstract The pursuit of new scientific research disciplines that design and implement human intelligence is no longer imagination or science fiction. Contemporary scientists worldwide use innovative ideas and wisdom to nurture this land of wisdom. They began to focus on robot design and implementation, integrating software systems and hardware devices to mimic human intellectual behavior through computational and artificial intelligence. This chapter explores the main theme of the book, Artificial Intelligence (AI). First, we begin with a general definition and its interpretation followed by a brief history review. Next, we analyze one of the most interesting AI experiments: The Turing Test. After that, we study two of AI's main themes strong AI and weak AI, and introduce five main components that perform the latest AI applications: machine learning, data mining, computer vision, natural language processing, and ontological-based search engines.

The pursuit of new scientific research disciplines that design and implement human intelligence is no longer imagination or science fiction since the Dartmouth conference. Contemporary scientists worldwide use innovative ideas and wisdom to nurture *this land of wisdom*. They began to focus on robot design and implementation, integrating software systems and hardware devices to mimic human intellectual behavior through computational intelligence and AI.

In this chapter, we will explore the main theme of the book, *AI*. First, we begin with a general definition and its interpretation followed by a brief history review. Next, we analyze one of the most interesting AI experiments: *The Turing Test*. After that, we study two of AI's main themes: *strong AI and weak AI*, and introduce five main components that perform the latest AI applications: machine learning, data mining, computer vision, natural language processing, and ontological-based search engines.

2.1   What is AI?

Artificial intelligence (AI) is an amazing term for most of us, not only in computer science, but also in other disciplines including neuroscience, psychology, and philosophy. Although there is no formal definition of AI, the definitions from the *American Association of Artificial Intelligence (AAAI)* and *Webster Dictionary* may be a good starting point (Lee 2006):

The scientific understanding of the mechanisms underlying thought and intelligent behavior and their embodiment in machine

American Association for Artificial Intelligence (AAAI)

The capacity of computers or programs to operate in ways to mimic human thought processes, such as reasoning and learning

Webster's New College Dictionary

Computer scientists Prof. Stuart Russell and Prof. Peter Norvig conducted a comprehensive survey of contemporary definitions in their excellent AI textbook *Artificial Intelligence: A Modern Approach* (Russell and Norvig 2016). They define AI in four main categories: (1) systems that think like humans; (2) systems with rational thinking; (3) systems that operate like humans; and (4) systems that operate rationally. It seems that the diversity of AI definitions is mainly reflected in two aspects: (1) the focus of functions is different and (2) different interpretations of *intelligence*. For (1), there are two controversial focal points: (a) behavior, that is, acting like a person; (b) cognitive style, that is, thinking like a person. For (2), most AI books and the literature focus on cognitive methods that attempt to interpret AI as a human rational thinking process imitations (Lee 2006). The reason is that they interpret intelligence as *rational thinking*. But is this really the case? In short: Are humans always rational species in all our behaviors and psychological processes?

The answer is simple, *no*. We often complete tasks in a relatively unreasonable (so-called *fuzzy*) way, select alternatives, and make decisions without even understanding the rationale behind. It is believed that such irrational behavior and decision-making are valuable assets of what we call *intelligence* (Fig. 2.1).

From the author's point of view, intelligence is the overall human ability to acquire, build, develop, and manipulate knowledge (i.e. our *thinking process*) simultaneously. *Thinking* can be classified as an intellectual process with three main categories:

1. *Logical thinking* (from derived knowledge),
2. *Lateral thinking* (from inspired knowledge), and
3. *Intuitive thinking* (from intuition).

Figure 2.2 illustrates the three levels of the thinking process.

Similarly, due to design and implementation (and systems) changes and diversification, AI should be able to imitate *human thoughts and behaviors*.

In the author's book *Fuzzy-Neuro Approach to Agent Applications* (2006), AI is defined as

The exemplification of human intellectual thoughts, acts, and behaviors for the design and implementation of intelligent systems, software objects (agents) and robotic systems (Lee 2006).

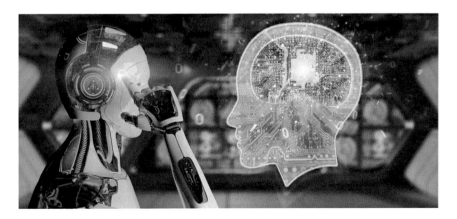

Fig. 2.1 Artificial intelligence (Tuchong 2020a)

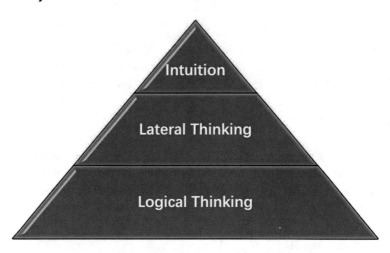

Fig. 2.2 Three levels of thinking process

2.2   A Brief History of AI

2.2.1   Pre-AI Stage (1943–1950)

The first scientific study before the famous *Dartmouth conference* should be by neurophysiologist Prof. Warren McCulloch (1898–1969) and mathematician Prof. Walter Pitts (1923–1969) with their remarkable research work *A logical calculus of the ideas immanent in nervous activity* published in *Bulletin of Mathematical Biophysics, 1943* (McCulloch and Pitts 1943). In this paper, they interpreted the neural activity in the human brain as a *logic gate* for information processing, the *biological neural network* we call nowadays.

Sir Alan Turing (1912–1952) published his influential paper *Computing Machinery and Intelligence* in *MIND*, 1950 (Turing 1950). He designed a test of a machine's ability to exhibit intelligent behavior that is equivalent to or indistinguishable from humans. Although it was not a mainstream scientific discipline at that time, the impact of the *Turing Test* coined the beginning of AI in human history.

2.2.2   First Golden Age of AI (1956–1974)

A group of young scientists working in logic and mathematics gathered on the campus of Dartmouth College in Hanover to discuss the possibility of producing a computer program that could *think*, in summer 1956. Their application was announced to the Rockefeller Foundation: *The study is to proceed on the basis of the conjecture that every aspect of learning or any other feature of intelligence can in principle be so precisely described that a machine can be made to simulate it* (Gardner 2008; McCorduck 2004). Professor John McCarthy (1927–2011), a mathematician and cognitive scientist at Dartmouth University, eventually became the founder and the first AI laboratory director of MIT and Stanford University (1963). He also coined *Artificial Intelligence (AI)* as a new discipline of scientific research. Founding members included cognitive scientist Prof. Marvin Minsky from Harvard (1927–2016) who eventually became the director of MIT AI laboratory, computer science and cognitive psychologist Prof. Allen Newell (1927–1992), and economist, political scientist, and cognitive psychologist Prof. Herbert Simon (1916–2001) from Carnegie Institute of Technology (now Carnegie Mellon College). Other researchers included Prof. Trenchard More, a mathematician and computer scientist from Princeton University; Prof. Arthur Samuel (1901–1990), IBM's pioneer in computer games and AI; Prof. Ray Solomonoff (1926–2009), an algorithmic probability inventor; and Prof. Oliver Selfridge, an AI pioneer (1926–2009) from Massachusetts Institute of Technology (2008). They introduced their latest research results and exchanged opinions on the possibility of developing machines that can imitate human behaviors and thinking process. Although there was no specific breakthrough at that time, this was a symbolic event in the birth of modern AI.

Professor Alan Newell, Prof. Herbert Simon, and system programmer John Clifford Shaw (1922–1991) developed the *Logic Theorist* between 1955 and 1956 (McCorduck 2004). This program proves 38 theorems of the first 52 theorems in Principia Mathematica of Whitehead and Russell, also known as the first AI program in human history.

Newell's early success in AI projects led to *General Problem Solvers* (GPS) study (Newell et al. 1959). Unlike *Logic Theorist*, the goal of GPS is to imitate the human ability to solve problems, the ultimate AI challenge. It turned out that within a limited range of problems handled, the sequence of the program took account of subgoals, and possible actions alike to human methods with the same problems. GPS cognitive model success led Newell and Simon (1976) to propose the famous physical symbol system hypothesis: *A physical symbol system has adequate and necessary means for general intelligent action.*

Emeritus Prof. Herbert Gelernter (1929–2015) established the *Geometry Theorem Prover* (GTP) (1959) to prove difficult geometry theorems. Artificial intelligence psychologist Prof. Frank Rosenblatt (Frank Rosenblatt) (1928–1971) proposed *perceptron,* an electronic device built on biology principles with learning ability in 1957. Its neural model became one of the neural network development foundations.

The US government and commercial organizations such as IBM attracted a large amount of research funding on AI research due to initial success between 1956 and 1974. Other major foundations included MYCIN introduction (Shortliffe 2012), an early backward linking expert system that uses AI to identify bacteria that caused severe infections, developed by Stanford University in the 1970s. However, because people have high expectations and expert systems had not yet been truly realized, the US government and commercial departments withdrew funds triggered during the *first winter of AI*.

2.2.3 First Winter of AI (1974–1980s)

The *First Winter of AI* between 1974 and 1980 was caused by the following factors:

Failure of Perceptron in 1969

The *perceptron* study (Bielecki 2018) invented by psychologist Prof. Frank Rosenblatt (1928–1971) predicted optimistically that perceptron may be able to learn, make decisions, and translate languages eventually. However, mainstream perceptron research ended abruptly in 1969, when cognitive scientist Prof. Marvin Minsky (1927–2016) and mathematician, computer scientist, and educator Prof. Seymour Papert (1928–2016) published a book *Perceptron* (Minsky and Papert 2017) outlined the limitations of the perceptron. Thereafter, the connectionist approach was abandoned in the next decade until Emeritus Prof. John Hopfield's work in the 1980s.

Lighthill Report in 1973

The UK Parliament asked Sir Michael James Lighthill (1924–1998) to assess the status and progress of AI in 1973. His famous *Lighthill report* criticized the failure of AI to achieve its grandiose objectives. He concluded that nothing can be done in AI, nor in other science fields. He also pointed out that many of AI's most successful algorithms would stall real-world problems and are suitable for solving toy versions only. The report led to the complete research disintegration in the UK, and US governments and commercial projects.

Major Funding Cut from DARPA in 1970s

The *Defense Advanced Research Projects Agency (also known as DARPA)* provided millions of dollars in AI research funding with a few strings attached in the 1960s. The situation that changed after the passage of the *Mansfield Amendment* in 1969 required DARPA to fund direct *mission-oriented research* instead of basic *nondirectional research*. Researchers found that results were difficult to achieve at that time and led to funding withdrawal.

2.2.4 Second Golden Age of AI (1980–1987)

WABOT-2 in 1980

Although AI experienced major setbacks in the West in the 1970s, it thrived in Japan. Japanese robot scientists have greatly promoted humanoid robots' development. Waseda University launched the *WABOT project* in 1967 and completed the world's first full-scale humanoid intelligent robot *WABOT-1* in 1972. Its limb control system made it possible to walk with tactile sensors and lower limbs, grasp and carry objects by hand. Its visual system allowed using external receptors such as artificial eyes and ears to measure distance and direction to objects. Its dialogue system communicated with humans in Japanese using the artificial mouth. All made it the first *android*. *WABOT-2* was produced at the same university in 1980 and has additional functions such as reading music pieces and playing the electronic organ.

XCON in 1980

XCON is an expert system developed at CMU in the 1980s. It saved the costs of Digital Equipment Corporation to approximately $40 million per year. Since then, more funds were obtained to support AI and expert systems such as LISP machine development.

Due to the success of the robot development, Japan launched the *fifth-generation computer project* in 1981 to manufacture AI machines (robots) to reason, translate languages, and understand music and pictures.

1982 was an important year for the *connectionism* concept in AI. Physicist and scientist Emeritus Prof. John Hopfield proposed his extraordinary Hopfield network (Coughlih and Baran 1995). It is a simple and surprising recurrent network with powerful memory storage and retrieval functions. Most importantly, Hopfield Network provides a model for understanding human memory with attractors in neural networks. The model was awarded the Dirac Medal of ICTP in 2002 for its interdisciplinary contribution in understanding biology as physical processes and calculations and granted him the Albert Einstein World Science Award in 2005. Professor Hopfield's main contribution to neural networks' development has injected innovative ideas including integrating adaptive learning into neural networks, studying neural network storage capabilities and associative storage network invention, and applying these networks to many complex problems including the famous *Traveling Salesman Problem* (TSP).

2.2.5   Second Winter of AI (1987–1993)

1987 Collapse of LISP Market

The dedicated AI hardware market collapsed in 1987. Workstation providers such as Sun Microsystems provided powerful alternatives to LISP machines, with companies like Lucid offering a LISP environment for this new type of workstation. General workstation performance became an increasingly difficult challenge for LISP machines. These companies and Franz LISP provided increasingly powerful LISP versions causing the market to collapse (Touretzky 2014).

Disappointment of the Fifth Generation Computer Projects

Japan's Ministry of International Trade and Industry has allocated US$850 million for the *fifth-generation computer project* (Scorrott, 2014) since the early 1980s. The goal was to build AI machines (robots) with human-like dialogue, that translate languages, interpret pictures, and with reasoning abilities. However, these goals were unable to be achieved until 1993, because expectations were much higher than other AI projects could achieve.

Underperformance of Expert Systems

The earliest successful expert system (Nikolopoulos 1997) such as XCON in early 1990 proved to be (1) too expensive for maintenance and support; (2) difficult to update changes; (3) no actual learning ability could be expected. For example, expert medical systems make serious mistakes on simple but unusual input. These expert systems have proved to be useful but have very few special contexts and limited fields.

Significant Funding Cut of AI Research

Expert system, robotics, and hardware system (such as LISP) development were below expectations alike to the first winter of AI, prompting governments from different countries (United States, Japan) and related agencies (such as DARPA) withdraw funding.

2.2.6   Third Golden Age of AI (1994–Now)

Birth of Internet and Intelligent Agents 1994 Onwards

A new paradigm called *intelligent agents* (Lee and Loia 2007) became widely recognized in the 1990s. This is different from the traditional computer system which consists of separate components such as I/O (input/output), process (programs and operations), and storage (hard disk and database). An intelligent agent consisted of an autonomous system that perceives its environment and takes actions to maximize its success chances. In other words, *intelligent agents* can be regarded as *software robots* consisting of both knowledge and operations, perceive, travel, and react to environments (and other agents) to complete its mission. In addition, the birth and booming of the Internet provide an excellent environment for the growth of *agent technology*.

IBM's Deep Blue Defeated World Chess Champion in 1997

IBM's Deep Blue AI system (Newborn 2013) became the first AI-based chess-playing system to defeat a reigning world chess champion, Garry Kasparov on May 11, 1997. Although it drew criticism for the lack of sufficient and revolutionary new paradigms, it was still an important milestone in AI computing over human intelligence.

Rises of AI Algorithms and Models in Various Practical Fields of Applications

AI scientists tried to develop specific AI algorithms and models instead of focusing on fully powerful intelligent robots to imitate human implementation, namely *microscopic AI* to imitate human intelligence in five major areas:

(1) Machine learning,
(2) Data mining,
(3) Computer vision,
(4) Natural language processing, and
(5) Ontological-based search engine.

The rapid Internet and mobile technology development led to many useful AI-based application and device releases such as smart watches to monitor users' daily health activities along with favorite song and movie searches from mobile apps in physical markets. We will study these in detail in the next part of the book.

2.3   Turing Test and AI

Everyone learned AI must have heard of the *Turing Test*. Why?

It is not only because it was invented by Sir Alan Turing (1912–1954), the father of theoretical computer science. Turing Test is truly an important test to demonstrate whether a computer system (or robot) has attained human's important abilities—*thinking* and *talking*.

The design and modeling of so-called *intelligent machines* had been studied prior to the Dartmouth conference. The most important and influential is the famous *Turing Test* named after computer pioneer Sir Alan Turing. Upon the completion of his remarkable work on the *Turing Machine* (Turing 1936, 1950) which became one of the algorithmic analysis foundational works for complex systems and problems, he started to focus on intelligent machines and theoretical research to investigate the relationship and differentiation between machine and human thoughts—the *Turing Test*.

The Turing test is a test of a machine's ability to exhibit intelligent behavior comparable to or indistinguishable from humans, developed in 1950. During the test, human judgement is based on natural language (such as English) between actual human and machine (robots) designed to produce human-like responses (Fig. 2.3).

The participants were separately seated in three different rooms and the judge was told that machine (robot) was one of the two participants. They used plain text channels (such as computer keyboards and screens) to communicate, usually with fixed time (approximately 15–20 min) to distinguish between the machine and human. If the judge was unable to distinguish between the machine and human at the end, the machine (robot) has passed the test.

Fig. 2.3 The Turing test (left: human; middle: judge; right: machine)

It was interesting to note from the test that the genius machine (robot) did not need to give all correct answers based on the judge's questions. On the contrary, it must answer them *naturally* and *humanly*. That is the subject of *human intelligence*. Since the first *Turing Test* commenced in 1950, scientists hinted that the computer would win the competition eventually by 2000. This test has always existed, no one dares to say that a person's machine can pass the test without any problems.

Although some criticized that these tests were unable to report machine intelligence faithfully, it was believed that *Turing Test* could reflect the main AI aspect, machine's *Natural Language Processing (NLP)* (Eisenstein 2019) capability only. However, the author concurs (Boden 1990) that natural language is not only a means of communication, but is also a vital, highly intellectual activity to reflect the test subject's overall wisdom.

2.4 Strong AI Versus Weak AI

There was even a controversy about AI general definition and classification since the Dartmouth meeting. An influential AI scholar Emeritus Prof. Robert Wilensky (1951–2013) pointed out in his book *Planning and Understanding: A Computing Method for Human Reasoning* that:

Artificial Intelligence is a field renowned for its lack of consensus on fundamental issues.

Robert Wilensky, 1983

A typical example is a debate between *strong (hard) AI* and *weak (soft) AI*. Some scientists believe that AI should focus on system or program design and implementation to imitate or simulate human thinking and behavior, the so-called *weak AI*. Others believe that AI systems should not only think and act like humans, but also think and act like humans consciously, the so-called *strong AI*. Turing test is a typical example of strong AI.

Sometimes, it is difficult to announce whether human-made decisions and choices are based on thinking, let alone about machines. For example, in chess games, it is usually thought that human activities involve mental thinking and intelligence. We often hear that world-class chess players do not use conventional *(logical) thinking* for decisions at most critical moments, but *intuition* for choices in other areas. In another extreme case, a computer program made decisions using breadth-first search only to play chess, making it difficult to classify whether the program really thinks. According to the author's previous intelligence and thinking definition, all the above situations can be defined as *intellectual activities*. They have different thinking ways and levels. Intuition itself is one kind of thinking (*intuitive thinking*), and breadth-first search should be another kind of thinking (*derived thinking*). Strong AI is not only an intellectual activity, but also has self-awareness thinking from an ontological perspective which will be studied in detail in the last part of the book (Fig. 2.4).

Fig. 2.4 Strong AI versus Weak AI (Tuchong 2020b)

2.4.1 Strong (Hard) AI

Strong AI (Nuncio 2019) that could perform any intellectual task successfully alike to humans refers to machine intelligence study. It also called *Artificial General Intelligence* (AGI). Although there is no strict definition, there is a wide agreement among AI researchers that intelligence is required to

- Reason by strategies, solve puzzles, and make judgments under uncertainty;
- Represent knowledge including common-sense knowledge;
- Plan;
- Learn;
- Communicate with natural languages, i.e. English; and integrate all these skills toward common goals.

Core Strong AI research areas include

- Building machines with intelligence, robotics in particular;
- R&D on generalized rule-based and case-based systems, e.g. IBM's Deep Blue on chess playing;
- R&D on consciousness, objective thoughts, and self-awareness.

Strong AI progress was sluggish due to complexities between 1950s and 1980s (Fig. 2.5).

2.4.2 Weak (Soft) AI

The emergence of software engineering, software (programs) design that can be executed by a computer (systems) on problem-solving supplies a new AI R&D area which is not focused on *General Intelligence* implementation, but more readily on mimicking human behaviors, particularly *thinking* and *problem-solving*, so-called *Soft AI* (or *weak/microscopic AI*).

Major Soft AI R&D includes (Konar 2018; Lee 2006).

- *Artificial neural networks* (ANN)—using computer models to mimic the human brain (neural) structure on thinking and problem-solving (Aggarwal 2018).
- *Genetic algorithms* (GA)—using computer models to mimic human genetic evolution, also called *evolutionary computing* (EC) or *evolutionary programming* (EP) (Michalewicz 1996).

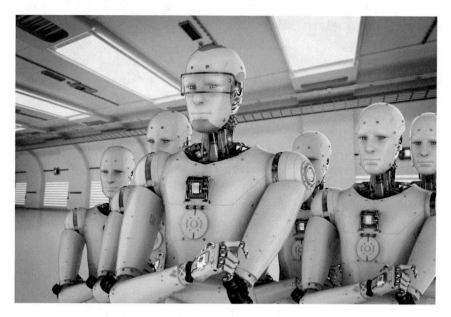

Fig. 2.5 AI Robotics (Tuchong 2020c)

- *Fuzzy logic* (or *fuzzy logic system, FLS*)—using computer models to mimic human imprecise (so-called *fuzzy*) matters and event determination, typical applications on real-time control systems (Belohlavek et al 2017) (Fig. 2.6).

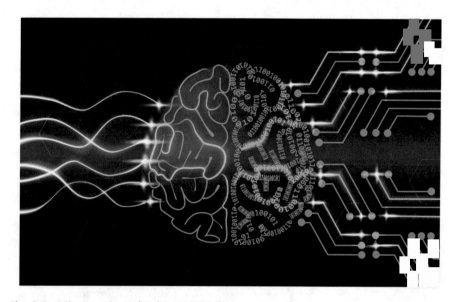

Fig. 2.6 Microscopic AI (Tuchong 2020d)

As one might see, the boundary between hard and soft AI becomes a blur. Modern AI becomes integrated disciplines nowadays including cognitive science, neuroscience, biological and artificial neural networks, evolutionary computing technology, robotic (engineering), data mining and deep learning, computer vision, natural language processing (NLP), etc.

2.5   Main Components of AI

Computing technology becomes state of the art in terms of (1) computational power, (2) memory and storage capability together with the Internet boom, mobile technology, and social media popularity.

Current AI development focuses on FIVE main areas of human dissimilation:

- Human learning processing—*machine learning* (ML) (Alpaydin 2016).
- Human thinking processing—*data mining* (DM) (Han et al. 2011).
- Human vision—*computer vision* (CV) (Davies 2017).
- Human language and conversations—*natural language processing* (NLP) (Eisenstein 2019).
- Human knowledge—*ontological-based search engine* (OSE) (Lim et al. 2011).

Figure 2.7 shows AI's main components.

2.6   Case Study—John Searle's Chinese Room Thought Experiment

Philosopher Prof. John Searle is noted widely for his philosophy of language, philosophy of mind, and social philosophy contributions. In his paper, *Minds, Brains, and Programs* published in *Behavioral and Brain Sciences* in 1980, he devised a thought experiment, the so-called *Chinese Room Thought Experiment* (Preston and Bishop 2002), to argue that a digital computer was unable to manifest a *mind*, lacked *understanding* or *consciousness* at program performance, regardless of how intelligent or human-like the program behaves.

The *Chinese Room Thought Experiment* is like this:

The *rule-of-game* was based on the Turing Test, this time it used the Chinese language—with Chinese characters. The machine was replaced by

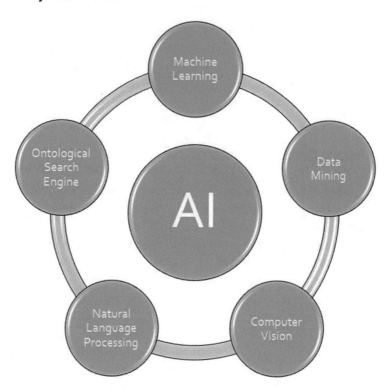

Fig. 2.7 AI's main components

a comprehensive translation book that could translate any Chinese characters and phrases to English and respond back in Chinese language. In other words, it is a kind of Chinese–English translation program (or system). The argument was that a computer system could perform the complex translation and respond perfectly (to deceive the judge) but it didn't imply that it has intelligence (Fig. 2.8).

Based on the same argument, he believed that the Turing Test was unable to determine whether a computer program (robot) is intelligent or not. What do you think?

2.7   Conclusion

AI is different from other disciplines such as philosophy, history and natural science like physics and chemistry. In this chapter, we study AI's general definition and its vicissitudes of the past 70 years.

There were several possible reasons that caused to such outcomes:

Fig. 2.8 Chinese room thought experiment

1. Over expectation
   The two *AI Winters* attributed to one major factor—over expectation. The fact that many AI-based systems such as expert system XCON had a principal defect that it was highly sensitive to inputs and domain. In other words, even though it is actually for a specific domain, it failed to perform with unusual inputs. The same occurred in other AI and expert systems.
2. Challenge of ultimate knowledge—Human intelligence
   Unlike other disciplines based on fundamental grounds and models, AI challenges the ultimate knowledge of human civilization—human intelligence.
3. Computer technology
   AI began from neural network research in the 1940s but computational power was limited which hindered its development significantly at that time. The same occurred in the second AI winter which was due to insufficient CPU processing power and data storage capability on expert system and robotic. CPU processing power and big data improvement set out the success of these systems.
4. Funding sources
   Fund withdrawal from governments and commercial institutions was due to projects and applications being unable to meet the requirements. However, AI research nowadays are required to be realistic, driven by end users, commercial startups, and markets' viewpoint.

References

Aggarwal, C. C. (2018). *Neural networks and deep learning: A textbook*. Springer.

Alpaydin, E. (2016). *Machine learning: The new AI* (MIT Press Essential Knowledge series). The MIT Press.

Belohlavek, R., et al. (2017). *Fuzzy logic and mathematics: A historical perspective*. Oxford University Press.

Bielecki, A. (2018). *Models of neurons and perceptrons: Selected problems and challenges* (Studies in Computational Intelligence Book 770, Kindle edition). Springer.

Boden M. A. (Ed.). (1990). *The philosophy of artificial intelligence*. Oxford University Press.

Coughlih, J. P., & Baran, R. H. (1995). *Neural computation in Hopfield networks and Boltzmann machines*. University of Delaware Publication.

Davies, E. R. (2017). *Computer vision: Principles, algorithms, applications, learning* (5th ed.). Academic Press.

Eisenstein, J. (2019). *Introduction to natural language processing* (Adaptive Computation and Machine Learning series). The MIT Press.

Gardner, H. (2008). *The mind's new science: A history of the cognitive revolution*. Basic Books.

Gelernter, H. (1959, June 15–20). Realization of a geometry theorem proving machine. In *Proceedings of the International Conference Information Processing* (pp. 273–282), Paris.

Han, J., et al. (2011). *Data mining: Concepts and techniques* (The Morgan Kaufmann Series in Data Management Systems, 3rd ed.). Morgan Kaufmann.

Konar, A. (2018). *Artificial intelligence and soft computing: Behavioral and cognitive modeling of the human brain*. CRC Press.

Lee, R. S. T. (2006). *Fuzzy-neuro approach to agent applications: From the AI perspective to modern ontology*. New York; Berlin: Springer.

Lee, R. S. T., & Loia, V. (2007). *Computational intelligence for agent-based systems*. New York; Berlin: Springer.

Lim, E. H. Y., Liu, J. N. K., & Lee, R. S. T. (2011). *Knowledge seeker: Ontology modelling for information search and management: A compendium*. Berlin: Springer.

McCorduck, P. (2004). *Machines who think: A personal inquiry into the history and prospects of artificial intelligence* (2nd ed.). CRC Press.

McCulloch, W. S., & Pitts, W. (1943). A logical calculus of the ideas immanent in nervous activity. *Bulletin of Mathematical Biophysics, 5,* 115–133.

Michalewicz, Z. (1996). *Genetic algorithms + data structures = evolution programs*. Springer.

Minsky, M., & Papert, S. A. (2017). *Perceptrons*. MIT Press.

Newborn, M. (2013). *Deep blue: An artificial intelligence milestone* (Springer Textbook). Springer.

Newell, A., Shaw, J. C., & Simon, H. A. (1959). Report on a general problem-solving program. In *Proceedings of the International Conference on Information Processing* (pp. 256–264).

Newell, A., & Simon, H. (1976). Computer science as empirical inquiry: Symbols and search. *Communications of the ACM, 19*(3), 113–126.

Nikolopoulos, C. (1997). *Expert systems: Introduction to first and second generation and hybrid knowledge based systems*. CRC Press.

Nuncio, R. (2019). *Strong artificial intelligence: Understanding the AI revolution* (Kindle edition). Ajijic Books Publishing.

Preston, J., & Bishop, M. (2002). *Views into the Chinese room: New essays on Searle and artificial intelligence*. Clarendon Press.

Rosenblatt, F. (1958). The perceptron: A probabilistic model for information storage and organization in the brain, Cornell aeronautical laboratory. *Psychological Review, 65*(6), 386–408.

Russell, S., & Norvig, P. (2016). *Artificial intelligence: A modern approach* (3rd ed.). Prentice Hall.

Scorrott, G. G. (2014). *The fifth generation computer project: State of the art report*. Pergamon Publication.

Shortliffe, E. H. (2012). *Computer-based medical consultations: MYCIN* (Artificial intelligence series, kindle edition). Elsevier.

Touretzky, D. S. (2014). *Common LISP: A gentle introduction to symbolic computation* (Dover Books on Engineering). Dover Publications.

Tuchong. (2020a). *Artificial intelligence*. Retrieved May 3, 2020, from https://stock.tuchong.com/image?imageId=539155542644359533.

Tuchong. (2020b). *Biological neural networks*. Retrieved May 3, 2020, from https://stock.tuchong.com/image?imageId=539155551234293979.

Tuchong. (2020c). *AI robotics*. Retrieved May 3, 2020, from https://stock.tuchong.com/image?imageId=467402221274661149.

Tuchong. (2020d). *Microscopic AI*. Retrieved May 3, 2020, from https://stock.tuchong.com/image?imageId=427423866836615175.

Turing, A. (1936). On computable numbers, with an application to the Entscheidungs-problem. *Proceedings of the London Mathematical Society, 42*(2), 230–265.

Turing, A. (1950). Computing machinery and intelligence. *Mind, LIX, 236*, 433–460.

Wilensky, R. (1983). *Planning and understanding: A computational approach to human reasoning*. Addison-Wesley Publishing Company.

Part II

AI Technology

3

Machine Learning

Previously, we might use machine learning in a few sub-components of a system. Now we actually use machine learning to replace entire sets of systems, rather than trying to make a better machine learning model for each of the pieces.
Dr. Jeff Dean (Lead of Google AI team, born 1968)

Abstract This chapter explores the first and foremost AI component and technology—machine learning (ML). We explore how humans learn and three major machine learning models: supervised learning, unsupervised learning, and reinforcement learning. We also study how the human brain works to think and learn—biological neural networks which lead to the design of artificial neural networks (ANN)—mathematical and computational counterparts to simulate human memory, learning, and thinking processes. To illustrate different types and models of memory storage, thinking and learning operations, we introduce four basic types of ANNs: (1) Auto-associative network for associative learning; (2) Hopfield network for memory storage and retrieval; (3) Feedforward backpropagation neural network for supervised learning; and (4) Actor-critic multi-agent-based model for reinforcement learning.

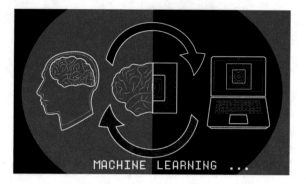

Fig. 3.1 Machine Learning (Tuchong 2020a)

Algorithms and models pertaining to microscopic AI cover five major components:

- Machine learning (ML)
- Data mining (DM)
- Computer vision (CV)
- Natural language processing (NLP)
- Ontological-based search engine (OSE).

We begin with machine learning (ML).

Machine learning is the foremost and fundamental AI component.

Many AI scientists, the author included, believe that any good AI application should inherit certain machine learning skills or capabilities; it should NOT claim to be an intelligent system if otherwise.

In other words, a sophisticated chess-playing system (robot) without machine learning skills can only be regarded as ordinary computer system without any intelligence, let's alone to be an *expert system* (Fig. 3.1).

So, what is Machine Learning?

3.1   The von Neumann Machine

Anyone who took computer fundamental courses knew the basic architecture of the computer system comes from *von Neumann Architecture*, also known as the *von Neumann Model*. It was a computer architecture proposed by polymath Prof. John von Neumann (1903–1957) in 1945 (Rojas and Hashagen 2000) based on Sir Alan Turing's work on *Turing Machine*. The design was published in a document called *First Draft of a Report on the EDVAC*. It described the first stored-program computer. The early computers such as ENIAC were hard-wired to do a single task. If the computer had to perform a different task, it required to rewire with a tedious process. A stored-program computer is a general-purpose computer built to run different programs. The *von Neumann Machine* can be considered a prototype of the modern computer system.

Figure 3.2 illustrates the basic components of *von Neumann Machine* which consists of

- Central processing unit (CPU)—system *brain* consists of system control unit, basic arithmetic, and logical unit.
- Memory unit—data and information *storage*.
- Input and output units—data input (e.g. keyboard) and output (e.g. screen).

Computer systems, regardless of speed or network, vary among supercomputers, mainframes, workstations, PCs, and mobile devices we use today; their basic architectures are all inherited from the blueprint of *von Neumann Machine*.

Are they *intelligent machines*?

Why or why not?

Let's take a look at two scenarios of computer systems applying to our daily activities.

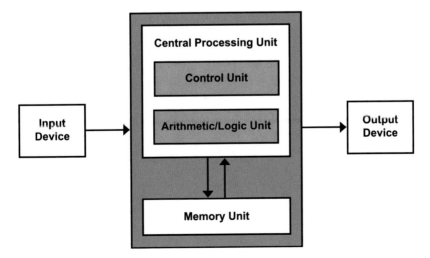

Fig. 3.2 Basic architecture of *von Neumann Machine*

3.2   Case 1–7—Day Weather Forecast System

First, look at a typical *7-Day Weather Forecast System.*

It makes no difference where you live, we all know what the weather forecast is. In fact, one of the major computer system applications (super-computer system, in particular) is *weather forecasting.*

Major weather forecast systems at present are called *Numerical Weather Prediction (NWP)* (Kalnay 2003). A typical NWP uses weather observations (such as temperature, air pressure, wind speed, and directions) of global grid points from different atmospheric levels as initial conditions. Numerical weather prediction is calculated by using fluid dynamics, thermodynamics equations, and Newton's laws of motion equations to forecast weather components such as air pressure, temperature, humidity, wind speed, and directions of every grid point for the future 168 h (7 days), with an overall accuracy of over 80–85%. Figure 3.3 shows a typical 7-day weather forecast results.

This forecast system is complex and powerful without argument, and it is a very typical *von Neumann Machine.*

But is there *Intelligence*? Can it *learn*? Can it *improve by itself*?

The answer is certainty *no.*

Fig. 3.3 7-day weather forecast system (Tuchong 2020b)

This system *crushes* with global grid point inputs, runs the forecast program, and produces forecast results regardless of its sophistication. The machine itself never *learns* or becomes *smarter* even though it has forecasts for years.

3.3   Case 2—AlphaGO

AlphaGo (DeepMind 2020) is an AI-based computer program that plays the board game Go. AlphaGo was developed by DeepMind Technologies in 2015 and acquired by Google. *Go* is an abstract strategy board game invented by the Chinese over 2500 years ago. It is believed to be the most complicated board game in human history.

The Go game (Pumperla and Ferguson 2019) is played by 2 players. The chess pieces are called *stones*. One player uses white stones and the other uses black. The players take turns placing the stones on vacant intersection points of the chessboard. Once placed on the board, the stones may not be moved, but are to be removed from the board if *captured*. Capture occurs when a stone or group of stones is/are surrounded by opposing stones on all adjacent orthogonal points. The game proceeds until neither player wishes to make another move. When a game concludes, the winner is determined by counting each player's surrounded territory along with the captured stones.

It is different from the standard *von Neumann Machine* where an AI-based chess-playing system has a unique feature such that it can *learn* from the game.

In other words, just like playing a game with an experienced *chess master*. When we play chess each time, the chess master will use his/her experience and knowledge to a greater extent. The most important fact is that he/she will improve his/her own chess-playing strategies to learn and predict our moves, to revise his/her own strategies and increase winning possibilities.

This is what we call *machine learning—intelligence* (Fig. 3.4).

3.4   How Humans Learn?

What is *Machine Learning*?

To answer this question, let's begin with *How can humans learn?*

Maybe it is perplexing to think about the approach.

Let's think of how we train children to learn. For example, matching a simple shape with a name as in Fig. 3.5.

Fig. 3.4 AI-based Go game (Tuchong 2020c)

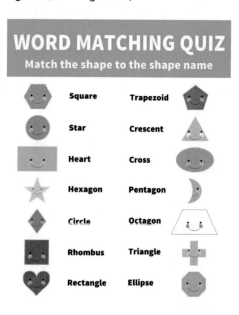

Fig. 3.5 Word matching quiz using associative learning (Tuchong 2020d)

The method is simple but is commonly used by humans to learn an *associ-ating* concept/idea (e.g. shape) with another concept/idea (e.g. shape name). We call this learning method *associative learning*.

Frankly, *associative learning* (Wills 2005) is a distinctly important learning method that occupies almost 80% of our entire learning life which includes

- Languages;
- Translation;
- Pattern and objective association;
- Concept and idea association.

Latest cognitive science believes that most low-level sensations such as touch, smell, and vision are all related to *associative learning*.

So, one might wonder: *What are the characteristics of this type of learning?*

This type of machine learning is characterized generally as *learning by example*.

3.5   Three Pillars of Machine Learning Methods

There are many different types of learning methods of our lives (Alpaydin 2016) that can be categorized into three major methods as illustrated in Fig. 3.6:

- Supervised learning,
- Unsupervised learning, and
- Reinforcement learning.

These learning methods consolidate with all various types of learning objectives and activities which include

- Pattern association and recognition (Bishop 2006),
- Object (2D, 3D) recognition and vision (Davies 2017),
- Concept learning and memorization (Gluck et al. 2016),
- Logical thinking and decision-making (Dettmer 2007), and
- High-level mental thinking and deduction process (Haber 2020).

It is important that although these learning processes are separate issues technically, humans combine them throughout their overall daily learning,

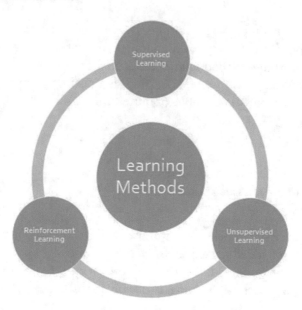

Fig. 3.6 3 pillars of machine learning methods

knowledge memory, and recalling processes subconsciously. Contemporary neuroscience (McNamara 2019) even believes humans sleep and dream not only to regain health and conscious mind, but more importantly to provide mechanisms for our brains' neural networks to reorganize memory and knowledge we experienced and learnt during interactions with the environment in the daytime.

3.6   Supervised Learning

Supervised Learning (Alpaydin 2016) is the basic type of machine learning method. It is characterized by learning of *example + outcome* in the sense that no matter what the kind of learning objective is (e.g. pattern matching and memory of words), the learning is performed by presenting *sample input + target output* pairs.

Typical examples are

- Pattern association with words.
- Concept association with meanings.
- Concept and meaning memorization.
- Logical deduction (if then … else).
- Logical induction.

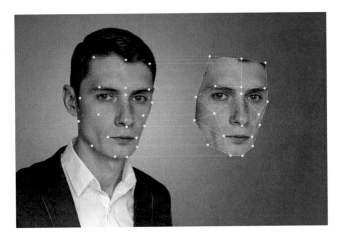

Fig. 3.7 Human face recognition with supervised learning method (Tuchong 2020e)

Let's use human face recognition as an example. As we know, human is competent to recognize human faces. The question is: *How can we build a machine to mimic humans to recognize human faces?*

Technically speaking, human face recognition (Li and Jain 2005) is a very primitive and basic supervised learning activity. Without training our brain on how to memorize a specific human face as per its name (a kind of pattern–name association), how can we recognize a *familiar* human face on the street? Another important feature is that such activity is usually performed *subconsciously* (or we can say *hard-coded*) in our brains. For example, whenever we see anyone, our brain will try to recognize his/her with our memories to examine whether we know him/her within seconds.

Latest works on cognitive science and neuroscience revealed that human is the best to perform such amazing job because we can extract important features subconsciously from a human face using the so-called *facial landmarks* and their relationship altogether as displayed to recognize nose, eyes, mouth, and ears separately. These theories inspire AI R&D in facial recognition to be studied in the following chapter (Fig. 3.7).

3.7   Unsupervised Learning

Unsupervised learning (Celebi and Aydin 2016) is a type of machine learning that looks for previous undetected patterns in a dataset of no pre-existing labels with minimum human supervision which is in contrast to supervised learning that usually uses human-labeled data. It is also known as *self-organization* that allows probability density modeling over inputs. In other words, unsupervised learning can be regarded as a kind of *self-learning* characterized without any specific target(s) or correct answers.

Typical examples include

- Pattern clustering,
- Figure-ground segmentation, and
- High-level mental process.

3.7.1 Pattern Clustering

Pattern clustering (Hennig et al. 2015), or *clustering*, is the task of grouping a set of objects situated in the same group (called a *cluster*) that are more similar (or close) to each other than to those in other groups (clusters). Pattern clustering includes object and pattern clustering and the auto-grouping process. It is believed that the inherited and natural process of human learning capacity can be achieved without prior knowledge. Apart from visual pattern clustering, nonvisual and mental clustering concepts and ideas also fall into this learning category.

Figure 3.8 shows a typical example of dots patch pattern clustering into two distinct clusters. In the coming section, we will reveal how we can train our brains to perform this process.

Fig. 3.8 Example of pattern clustering into two clusters

3.7.2 Figure-Ground Segmentation

Figure-ground segmentation explores the way how we differentiate an object from its background. It has more fundamental visual and unsupervised learning realized subconsciously in our minds. Figure 3.9 shows the famous visual psychology Rubin-vase experiment, the so-called *Gestalt psychology* (Koffka 2014). Gestalt psychologists emphasize that organisms perceive entire patterns or configurations, not merely on individual components. In this experiment, Gestalt psychologists maintain that human can either identify a vase or two human face profiles but not two items simultaneously.

It is different from the recognition of an object such as human face mentioned in supervised learning. This learning machine must have a set of objects' databank (e.g. human face databank in face recognition problem) to learn or compare with. Figure-ground segmentation is considered as a typical kind of unsupervised learning in machine learning where there should be no target-object to match or compare with. In other words, a typical figure-ground segmentation engine can automatically *extract* an object from the scene (background) before it *recognizes* what the object is.

Fig. 3.9 Rubin-vase experiment in Gestalt visual psychology

3.7.3 High-Level Mental Process

High-level mental process such as *book reading* is a kind of unsupervised learning in the sense that during the process, our mind tries to *conceptualize* words, clauses, and passages into *concept* and *idea clusters*, compares with what we had learnt from memory, and stores them subconsciously as *new knowledge*. In many typical cases, unsupervised learning is performed in our mind subconsciously and automatically.

Many primitive visual psychologies including Gestalt psychology are highly related to human-inherited unsupervised visual learning capability. The latest psychology development also reveals that unsupervised learning constitutes a major human development psychology component in children and adolescence learning behaviors.

In unsupervised learning, an AI system is presented with unlabeled, uncategorized data, and system's algorithms act on data without prior training. The output is dependent upon coded algorithms. Subjecting a system to unsupervised learning is a way of AI testing. Unsupervised learning algorithms can perform more complex processing tasks than supervised learning systems. Yet unsupervised learning can be more unpredictable than the alternate model. An unsupervised learning AI system might, for example, figure out on its own how to sort cats from dogs; it might also add on unforeseen and undesired categories to deal with unusual breeds, creating a clutter instead of an order (Fig. 3.10).

Fig. 3.10 Example of high-level mental process, i.e. book reading (Tuchong 2020f)

3.8   Reinforcement Learning

Reinforcement learning is also known as *conditioning* (Domjan 2016) in *Behavioral Psychology*. It is a type of learning and training process that will strengthen an organism's future behavior whenever that behavior is preceded by a specific antecedent stimulus. Laboratory research on reinforcement is usually dated from the work of psychologist Prof. Edward Thorndike (1847–1949), known for his experiments with cats escaping from *Puzzle Boxes* in 1929.

Psychologist and behaviorist Prof. B. F. Skinner (1904–1990) is referred to as the father of *Operant Conditioning*, and his work is frequently cited in connection with this topic (Skinner 2014). His 1938 book *The Behavior of Organisms: An Experimental Analysis* (Skinner 2006) initiated his lifelong study of operant conditioning and its application to human and animal behaviors. Following the ideas of physicist Prof. Ernst Mach (1838–1916), Skinner rejected Thorndike's reference to unobservable mental states such as satisfaction, building his analysis on observable behavior and its equally observable consequences. He believed that classical conditioning was too simplistic for use to describe something complex as the human behavior. Operant conditioning, in his opinion, was more suitable to describe human behavior as it examined intentional behavior causes and effects.

He developed a device called the *Skinner Box* which records each response provided by an animal and its unique reinforcement schedule that was assigned. Pigeons are placed in operant conditioning chambers and receive a food pellet for pecking at a response key. Some pigeons receive a pellet for every response (continuous reinforcement) while others obtain a pellet only after a certain amount of time or numbers of response occurred as in Fig. 3.11.

There are two types of *reinforcement learning*, known as positive reinforcement by *rewards* and negative reinforcement by *punishments*. If the object under investigation performs the desirable action or response, a *reward* is offered, e.g. food, whereas if the object performs unwanted behavior, *punishment* action is taken, e.g. no food.

Tremendous research on reinforcement learning have been done for the past half century, ranging from animal trainings to human learning memory enhancement, which is an important research domain in behavioral and psychological development.

The entire Western education exerts reinforcement training for students to awards on good examination grades or fair punishment on poor examination

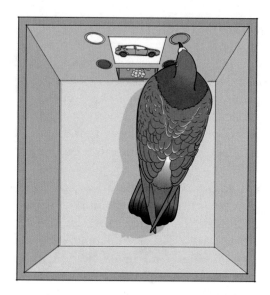

Fig. 3.11 Skinner box test for reinforcement learning using pigeon as a test subject

grades to trigger and reinforce learning *incentives*. The study of reinforcement training and learning had produced an enormous body of reproducible experimental results.

Reinforcement in the business world is essential in driving productivity. Employees are constantly motivated by positive stimuli such as promotion and bonus. Employees are also driven by negative reinforcement such as work on off days if weekly workloads are not completed on time.

Even today, reinforcement training and learning is a central concept and procedure in special education such as applied behavior analysis. The experimental behavior analysis is also a core concept in certain medical and psychopharmacology models such as addiction, dependence, and compulsion, and of course; a major research topic in AI and machine learning (Fig. 3.12).

3.9 Biological Neural Networks

3.9.1 Our Brain

How can we Learn and Think?—our brain.

Artificial Neural Networks (Aggarwal 2018), *ANN* (or *Neural Networks* in short), is a microscopic AI major component that focuses on studying and

(a)   dolphin training   (b)   dog training

Fig. 3.12 Reinforcement training for dolphins and dogs (Tuchong 2020g, h)

modeling of an intelligent system to mimic one of the most important human organs—*our brain.*

The first scientist to work on *brain science* area was biologist and pathologist Prof. Camillo Golgi (1843–1926) who invented the staining method to investigate neural activities inside the brain. By using this method, he proposed that the brain is made up of syncytium—a sponge-like tissue that is activated by the staining operation. Based on his discovery, neuroanatomist Prof. Santiago Ramón y Cajal (1852–1934) proposed an innovative idea that *"these staining tissues were not sponge-like elements, but rather the collections of brain cells called neurons"*, which were interlinked together to form a complex group—*neural networks* (Mazzarello 2010). Figure 3.13a, b shows 3D modeling and graphical illustration of *biological neural networks* in our brain.

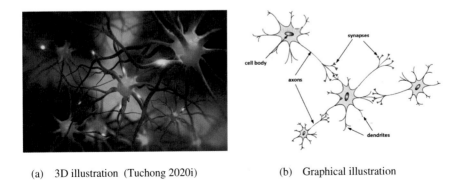

(a)   3D illustration (Tuchong 2020i)   (b)   Graphical illustration

Fig. 3.13 Biological neural networks in our brain

As shown, each neural cell (neuron) consists of

- *Nucleus*—central body of the neuron.
- *Axon*—prolonged filament connections to other neurons.
- *Dendrites*—tree-like structures branching from the neuron.
- *Synapse*—axon tips (junctions) contact with other neurons by attaching to the dendrites of neighboring neurons.

3.9.2 Integrate-and-Fire Operations in Biological Neural Network

Before 1943, almost all neuroscientists believed that the sole purpose of neurons was to process energy, but how they work, e.g. process information and store memory, remains unknown.

Neurophysiologist and cybernetician Prof. Warren McCulloch (1898–1969), and logician Prof. Walter Pitts (1923–1969) published an influential paper in 1943 *A logical calculus of the ideas immanent in nervous activity* (McCulloch and Pitts 1943) which triggered the birth of *artificial neural networks* (ANN).

They proposed that the main function of neural activities was to process information, not energy storage. The function of neurons was like *logical switches*. The signal transmission from one neuron to another at synapses is the result of a complex chemical process in which specific transmitter substances are released from junctions sending points. If the potential reaches a certain threshold, a pulse will be generated down the axon known as *firing* as in Fig. 3.14.

More importantly, they demonstrated how their proposed network (now called *artificial neural network*) could be used to perform basic logical operations such as *AND, OR, and NOT*.

This breakthrough not only solved the *century's mystery* of how biological neural network works, but also provided a solid foundation for digital computing technology development.

Although we now know that the neural activities in our brains are quite different from logical switches such as transistors in a digital computer, which are alike for nonlinear, and even chaotic *integrate-and-fire* operators for information transmission and processing, the discovery in 1943 coined the so-called *First Golden Age of Artificial Neural Networks*.

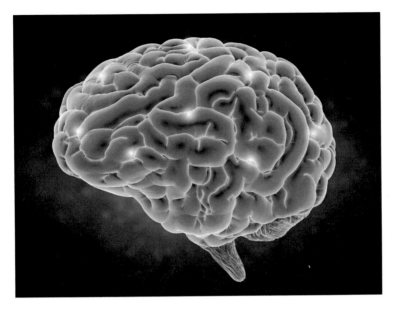

Fig. 3.14 Integrate-and-fire operations in biological neural network (Tuchong 2020j)

3.10 Artificial Neural Networks

3.10.1 A Neuron Model

As a direct analog of a biological neuron, the schematic diagram of neuron structure can be interpreted as a computational model in which synapses are represented by weights that modulate the effect of associated input signals, with the formulation given by (Fausett 1993; Aggarwal 2018):

$$y = f\left[\sum_{i=1}^{n} w_i x_i\right] \tag{3.1}$$

where x is the input signals, w's are the weights and y is the output.

The nonlinear characteristics exhibited by the neuron are represented by a transfer function $f(x)$ such as a binary or bipolar sigmoid function, given by.

Binary sigmoid function:

$$f(x) = \frac{1}{1 + e^{-\sigma x}} \tag{3.2}$$

Bipolar sigmoid function:

$$f(x) = \frac{1 - e^{-\sigma x}}{1 + e^{-\sigma x}} \tag{3.3}$$

where σ is the steepness parameter to control the curvature of the transfer function.

Figure 3.15a, b illustrates the contracts between a biological neuron in our brain and an artificial neural network neuron model.

The learning capability of an artificial neuron is achieved by adjusting the weights in accordance with a predefined learning algorithm, usually in the form of

$$\Delta w_j = \alpha \sigma x_j \tag{3.4}$$

where α is the learning rate and σ is the learning momentum.

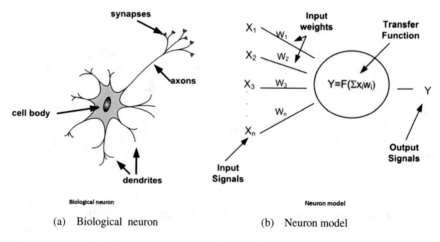

(a) Biological neuron   (b) Neuron model

Fig. 3.15 Biological neuron versus neuron model

3.10.2 Artificial Neural Network

Typical *artificial neural network (ANN)* consists of intermediate layer(s) known as *hidden layers* to facilitate nonlinear network system computational capabilities (Fausett 1993; Aggarwal 2018).

Classical ANNs such as the Feedforward Neural Network (FFNN) illustrated in Fig. 3.16 allow signals (information) to flow from input units to output units in a forward direction. As shown, a typical artificial neural network had a 3-layering structure that consists of (1) input layer with neurons x_i; (2) hidden layer with neurons y_j; and (3) output layer with neurons z_k. The feedforward neural network was the first and simplest type of artificial neural network devised. In this network, the information moves in only one direction, forward, from input nodes through hidden nodes (if any) and to output nodes.

Other basic ANNs include the classical *Kohonen self-organizing map (SOM)* and *Learning Vector Quantization* (LVQ) based on competition, and the *Adaptive Resonance Theory (ART)*, and of course, our main theme—*Feedforward Backpropagation Neural Network (FFBPN).*

ANNs can be regarded as multivariate nonlinear analytical tools known to be superior at recognizing patterns from noisy, complex data, and estimating their nonlinear relationships.

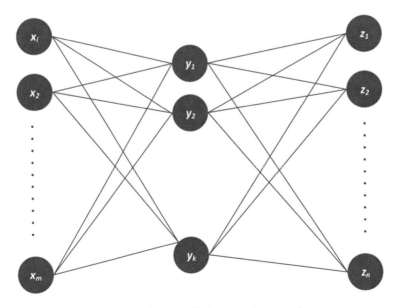

Fig. 3.16 Typical single hidden layer *artificial neural network*

Many studies revealed that ANNs had the distinguished capability to learn the underlying mechanics of time-series problems ranging from stock prediction and foreign exchange rates in various financial markets to weather forecasts.

3.10.3 Classification of Neural Networks by Machine Learning Technique

There were numerous neural networks proposed throughout the past half century with over 20 different artificial neural network (ANN) types commonly used (Lee 2006).

ANNs are commonly classified by (1) Machine learning technique and (2) Areas of application.

In terms of machine learning techniques, ANNs can be classified into three main categories as shown in Fig. 3.17:

(1) *Supervised Learning Neural Networks*

    Network learning (training) based on input–output (target output) pairs. Typical examples include *Feedforward Backpropagation Neural*

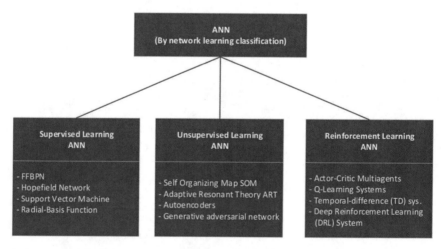

Fig. 3.17 Classification of neural networks by machine learning technique

Network (FFBPN), Hopfield Network, Support Vector Machine (SVM), Radial-basis Function Network (RBFN), etc.

(2) *Unsupervised Learning Neural Networks*

Neural networks that do not require any supervised learning and training strategies that include all kinds of self-organizing, self-clustering, and learning networks such as *SOM and ART (Adaptive Resonant Theory).*

(3) Reinforcement-Learning Neural Networks

Different from *supervised learning* (SL) with well-defined input–output pairs, *reinforcement-learning* (RL) trains neural networks with the adoption of feedback signals namely *reinforcement signal* (RS). With the right behavior, the network will respond with a positive RS to *award* the RL network, while toward the wrong behavior, the network will respond with a negative RS to *punish* the RL network. This method is particularly useful to tackle the optimization problem without exact target solutions such as trading strategy optimization.

3.11 Machine Learning Models

3.11.1 Associative Neural Network for Associative Learning

Associative learning (Findler 2016) is one of the most fundamental human intellectual behaviors to recall and handle memory storage revealed in Sect. 3.4. It is widely used by humans and machines for pattern recognition such as visual pattern identification and recognition for recalling human faces, voices, and music. It relates to knowledge and memory recalling association.

In general, an associative neural network (also known as *associative memory network* or *memory network* in short) is a single-layered neural network used to store a set of patterns (memory) for pattern association (or what we call *memory recalling*).

Figure 3.18 illustrates a typical associative neural network configuration.

Associative network training is conducted by iterative stored patterns presentation for weights updated according to the training algorithm.

Once the training is completed, the network can be used to associate not only stored pattern, but also the correct stored pattern upon an incomplete or noisy query pattern presentation.

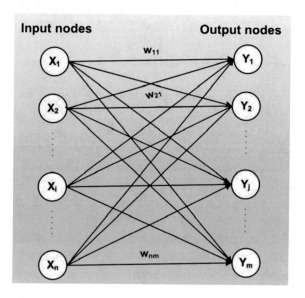

Fig. 3.18 Associative neural network

There are basically two major kinds of associative networks:

(1) *Auto-associative networks*—in which input (and query) patterns are the same type (and nature) as an associated pattern; and
(2) *Hetero-associative networks*—in which input (and query) patterns are completely different types (and nature) from associated patterns.

3.11.2 Hopfield Network for Memory Storage and Retrieval

In 1984, physicist and scientist Emeritus Prof. John Hopfield published his influential paper: *Neurons with graded response have collective computational properties like those of two-state neurons* (Hopfield 1984). He described how a simple recurrent auto-associative network can be used for content-addressable memory systems. This network can also be used for pattern recognition and tackle complex optimization problems such as the typical *traveling salesman problem* (TSP) (Applegate et al. 2007, Li et al. 2016).

The architecture of the Hopfield network is alike as a classical auto-associative network but with three basic differences:

(1) The Hopfield network is a recurrent network where output nodes are fed in one time-step as input in the next time-step.

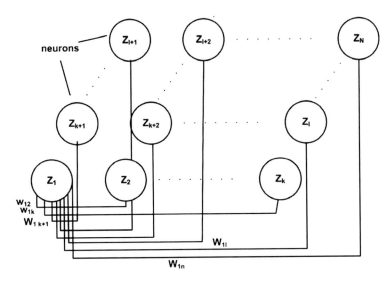

Fig. 3.19 Discrete Hopfield network model

(2) In classical associative network, all neurons will update their activations simultaneously, but in the Hopfield network only one neuron will be chosen to update its activation at a time and will then *broadcast* its new state to other members of the network.

(3) Each neuron will keep on receiving the *stimulus* from an external signal during the entire process.

Figure 3.19 illustrates the original system architecture of the discrete Hopfield network (Lee 2006).

One of the important points from the Hopfield network is that it demonstrates how a simple auto-associative network can be modified to produce a powerful memory storage and retrieval device. In fact, the vast application areas of Hopfield networks also triggered the rebirth of ANNs, and the exploration of how neural networks can be applied to complex problems in daily operations. Hopfield networks also provided a model for understanding human memory and important AI components (Lee 2006; Yang et al. 2017).

3.11.3 Feedforward Backpropagation Network for Supervised Learning

FFBPN provides a multilayer network architecture and is different from the previous two neural networks. A typical FFBPN consists of an input layer, a hidden layer, and an output layer. Although FFBPN can consist of several hidden layers, in most of the cases one hidden layer is usually sufficient (Fausett 1993; Aggarwal 2018).

FFBPN's network training consists of three main processes:

(1) The *feedforward* process of network training.
(2) The *error evaluation* process to calculate errors between calculated output values and target output values.
(3) The *backpropagation* errors process for weight adjustments.

Figure 3.20 illustrates the system architecture of a typical FFBPN. In the network architecture, *w's* denote the network weights between input and hidden layers, and *u's* denote the network weights between hidden and output layers. The total number of neurons in input, hidden, and output layers are n, t, and m, respectively. For the activation functions, a sigmoid function is adopted normally.

FFBPN is alike as most other neural networks, training stops when errors are bound within the tolerance level. As shown in the network training process, it learns by comparing network outputs with target outputs and adjusts the network weights between output layer/hidden layers and hidden layer/input layer in the backward mention. So FFBPN is particularly useful

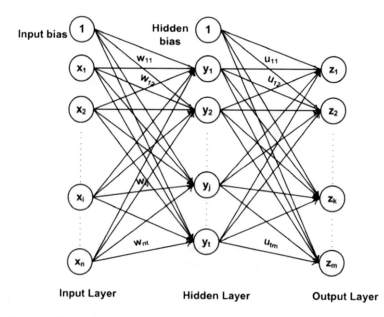

Fig. 3.20 Feedforward Backpropagation Network (FFBPN)

for Supervised Learning problems which have well-defined input/target-outputs for network training.

An FFBPN can model various kinds of pattern recognition problems such as character recognition, classification, and optimization. It can also be used at time-series prediction problems such as weather prediction and stock forecasting (Lee 2006).

3.11.4 Actor-Critic Multi-agent Model for Reinforcement Learning

There are many situations in which input/target-output pairs do not exist which are different from the supervised learning model with well-defined *"input/target-output"* pairs to train the network. *Reinforcement learning (RL)* is an area of machine learning concerned with how an AI system called *intelligent agents* (Lee and Loia 2007) ought to take actions in an environment in order to maximize the *rewards* (positive rewards) and minimize the *punishments* (negative rewards) alike to humans and other animals learnt from reinforcement training and conditioning.

RL does not need any *input/output pairs* for training. Instead, the focus is on finding a balance between *exploration* (of uncharted territory) and *exploitation* (of current knowledge) via *reward* and *punishment* mechanisms while the agent life cycle interacts with the environment. One of the most frequently used intelligent agent-based reinforcement learning systems is the *Actor-Critic Reinforcement Learning system (ACRLS)* as shown in Fig. 3.21.

A typical ACRLS consists of four components: *Environment Space (E), Action Space (A), Actor Agent (Actor), and Critic Agent (Critic).*

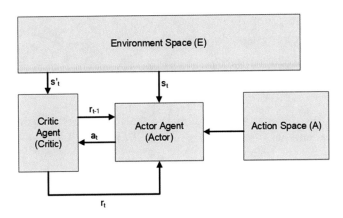

Fig. 3.21 Actor-critic reinforcement learning model

ACRLS visualizes its *world* as a collection of discrete-time-step states and actions. The role of the *Actor agent* is based on input signals provided at the current state at time t to respond with the *best* available actions (a), whereas the role of the *Critic agent* is to evaluate the *reward signal* (r) based on action(s) taken by the Actor and signals provided by new states at time t. Then the reward signal gives feedbacks as input to the Actor agent to decide the next action *a* at time t + 1.

RL is particularly useful to problems that involve a long-term versus short-term reward trade-off. Today, intelligent agent-based RL systems are frequently used in robot control and navigation, elevator scheduling, and telecommunications. It is also commonly found in AI-based board games such as backgammon, checkers, and GO (Wiering and Otterlo 2012; Sutton and Barto 2018).

3.12   Case Study—Learning at School

The best place to study learning methods initially, of course, is by *learning at school*. The 3 types of learning methods supervised learning, unsupervised learning, and reinforcement learning are always integrated and mixed with our daily learning activities that range from learning concepts and ideas from lectures, for tests, self-revision, and examinations.

The learning activities include

- Attending lectures;
- Participating in tutorials and labs;
- Reading books and papers;
- Completing assignments and exercises;
- Participating in group discussions;
- Completing term tests and quizzes;
- Study and revision;
- Attending the final exam.

Please identify the learning method(s) involved in each learning activity. Discuss how and why schooling consists of all kinds of learning activities (Fig. 3.22).

(a)   Reading & study   (b)   Test & examination

Fig. 3.22   Learning at school (Tuchong 2020k, l)

3.13 Conclusion

In this chapter, we explore the first and foremost AI component and technology—*machine learning* (ML). Since ML is a core AI characteristic, without it, the system is hardly an AI system.

We discover how humans learn. There are three major learning model components: supervised, unsupervised, and reinforcement learning. Although these three learning models are independent of each other, we normally use and combine them unconsciously for our daily learning and training operations.

We also studied how our brain works in order to *think* and *learn—biological neural networks* in our brain that lead to the design of artificial neural networks (ANN)—the mathematical and computational counterparts of the human brain to simulate human memory, learning, and thinking processes.

To illustrate different types and models of memory storage, thinking and learning operations, we introduce four basic types of ANNs:

- Auto-associative network for associative learning;
- Hopfield network for memory storage and retrieval;
- Feedforward backpropagation neural network for supervised learning;
- Actor-critic multi-agent-based model for reinforcement learning.

As one might see, all types of AI machine learning techniques are inspired by our knowledge and understanding of human thinking, learning processes, and memory storage which are *the soul of AI*.

References

Aggarwal, C. C. (2018). *Neural networks and deep learning: A textbook*. Springer.

Alpaydin, E. (2016). *Machine learning: The new AI* (MIT Press Essential Knowledge series). The MIT Press.

Applegate, D. L., et al. (2007). *The traveling salesman problem: A computational study* (Princeton Series in Applied Mathematics (40)). Princeton University Press.

Bishop, C. M. (2006). *Pattern recognition and machine learning* (Information Science and Statistics). Springer.

Celebi, M. E., & Aydin, K. (2016). *Unsupervised learning algorithms*. Springer.

DeepMind. (2020). *Alpha Go*. Retrieved May 6, 2020, from Official site: https://deepmind.com/research/case-studies/alphago-the-story-so-far.

Davies, E. R. (2017). *Computer vision: Principles, algorithms, applications, learning* (5th ed.). Academic Press.

Dettmer, H. W. (2007). *The logical thinking process: A systems approach to complex problem solving*. American Society for Quality.

Domjan, M. (2016). *The essentials of conditioning and learning* (4th ed.). American Psychological Association.

Fausett, L. V. (1993). *Fundamentals of neural networks: Architectures, algorithms and applications*. Pearson.

Findler, N. V. (2016). *Associative networks: Representation and use of knowledge by computers*. Academic Press.

Gluck, M. A., et al. (2016). *Learning and memory*. Worth Publishers.

Haber, J. (2020). *Critical thinking* (MIT Press Essential Knowledge series, kindle edition). The MIT Press.

Hennig, C., et al. (2015). *Handbook of cluster analysis* (Chapman & Hall/CRC Handbooks of Modern Statistical Methods). Chapman and Hall/CRC Press.

Hopfield, J. J. (1984). Neurons with graded response have collective computational properties like those of two-state neurons. *Proceedings of the National Academy of Sciences, 81,* 3088–3092.

Kalnay, E. (2003). *Atmospheric modeling, data assimilation and predictability*. Cambridge University Press.

Koffka, K. (2014). *Principles of Gestalt psychology*. Mimesis International.

Lee, R. S. T. (2006). *Fuzzy-neuro approach to agent applications: From the AI perspective to modern ontology*. New York; Berlin: Springer.

Lee, R. S. T., & Loia, V. (2007). *Computational intelligence for agent-based systems*. New York; Berlin: Springer.

Li, R., Qiao, J., & Li, W. (2016). A modified Hopfield neural network for solving TSP problem. In *Proceedings of the 12th World Congress on Intelligent Control and Automation (WCICA)* (pp. 1775–1780).

Li, S. Z., & Jain, A. K. (2005). *Handbook of face recognition*. Springer.

Mazzarello, P. (2010). *Golgi: A biography of the founder of modern neuroscience*. (A. Badiani & H. A. Buchtel, Trans.). New York: Oxford University Press.

McNamara, P. (2019). *The neuroscience of sleep and dreams* (Cambridge Fundamentals of Neuroscience in Psychology). Cambridge University Press.

McCulloch, W. S., & Pitts, W. (1943). A logical calculus of the ideas immanent in nervous activity. *Bulletin of Mathematical Biophysics, 5,* 115–133.

Pumperla, M., & Ferguson, K. (2019). *Deep learning and the game of Go.* Manning Publications.

Rojas, R., & Hashagen, U. (Eds.). (2000). *The first computers: History and architectures.* MIT Press.

Skinner, B. F. (2014). *Contingencies of reinforcement: A theoretical analysis* (B. F. Skinner Reprint Series; Edited by Julie S. Vargas Book 3). B. F. Skinner Foundation.

Skinner, B. F. (2006). *Behavior of organisms.* Copley Publishing Group.

Sutton, R. S., & Barto, A. G. (2018). *Reinforcement Learning: An introduction* (Adaptive Computation and Machine Learning series, kindle edition). A Bradford Book.

Tuchong. (2020a). *Machine learning.* Retrieved May 6, 2020, from https://stock.tuchong.com/image?imageId=423728176325853187.

Tuchong. (2020b). *7-day weather forecast system.* Retrieved May 6, 2020, from https://stock.tuchong.com/image?imageId=429156508183429378.

Tuchong. (2020c). *AI-based Go Game.* Retrieved May 6, 2020, from https://stock.tuchong.com/image?imageId=455145303749427362.

Tuchong. (2020d). *Word matching quiz using associative learning.* Retrieved May 6, 2020, from https://stock.tuchong.com/image?imageId=460465101781860718.

Tuchong. (2020e). *Human face recognition.* Retrieved May 6, 2020, from https://stock.tuchong.com/image?imageId=255770877421158487.

Tuchong. (2020f). *Example of high-level mental process such as reading books.* Retrieved May 6, 2020, from https://stock.tuchong.com/image?imageId=428989330876399970.

Tuchong. (2020g). *Reinforcement training for Dolphin training.* Retrieved May 6, 2020, from https://stock.tuchong.com/image?imageId=239306755947102314.

Tuchong. (2020h). *Reinforcement training for Dog training.* Retrieved May 6, 2020, from https://stock.tuchong.com/image?imageId=258964683821941024.

Tuchong. (2020i). *3D illustration of biological neural networks.* Retrieved May 6, 2020, from https://stock.tuchong.com/image?imageId=483388184041619483.

Tuchong. (2020j). *Integrate-and-fire operations in biological neural network.* Retrieved May 6, 2020, from https://stock.tuchong.com/image?imageId=260174455850139720.

Tuchong. (2020k). *Reading & study.* Retrieved May 6, 2020, from https://stock.tuchong.com/image?imageId=426568721668112651.

Tuchong. (2020l). *Test & examination.* Retrieved May 6, 2020, from https://stock.tuchong.com/image?imageId=313714891096850447.

Wiering, M., & Otterlo M. (2012). *Reinforcement learning: State-of-the-art* (Adaptation, Learning, and Optimization Book 12). Springer.

Wills, A. J. (2005). *New directions in human associative learning*. Psychology Press.

Yang, J., Wang, L., Wang, Y., & Guo, T. (2017). A novel memristive Hopfield neural network with application in associative memory. *Neurocomputing, 227,* 142–148.

4

Data Mining

Data is the fabric of the modern world: just like we walk down pavements, so we trace routes through data, and build knowledge and products out of it.
Prof. Ben Goldacre (Physician, born 1974)

Abstract This chapter explores various methods and technologies that involve data mining that includes KNN for clustering, decision tree for decision-making, regression for forecast and projection, and association rule for mining useful patterns. We also introduce deep neural networks (DNN) in data mining. The truth is there are many other useful data mining tools and technologies; the focus of the chapter is to provide readers an overview of concepts and key technologies, and more importantly, on how it can be applied to our real-world daily activities. A practical solution should involve data mining and knowledge discovery from different data sources, data formats, and appearances.

In Chap. 3, we learnt different types of machine learning methods and technologies ranging from pattern recognition, memory storage to forecasting to solve various real-world problems.

Is there any alternative to *see* the problem?

The answer is a definitely *yes—Data Mining*.

We refer to Kant's remarkable work *Critiques of Pure Reason* and interpreted it in layman terms: *What we perceive in our world depends on the way we see the world. The way we see our world will affect the way we solve the problem happening in our perceived world.*

In other words, if we consider a problem as a machine learning problem, what we can do (and will do) is to apply appropriate machine learning techniques to solve it. The question is: Is that always the case? Certainly not.

However, most of the real-world problems involve data and information. We believe by analyzing and exploring data, we may be able to extract some useful information or even knowledge for a solution. That is the spirit of *data mining*.

In this chapter, we will study different classical (non-AI) data mining methods and technologies, together with the latest AI-based data mining technology and how to apply it to our daily activities.

4.1   What is Data Mining?

Data mining (DM) (Tan et al. 2018) is the process of discovering patterns in large datasets involving methods at the intersection of machine learning, statistics, and database systems. DM is an interdisciplinary subfield of computer science and statistics with an overall goal to extract information (with intelligent methods) from a dataset and transform the information into a comprehensible structure for use.

Data mining is an important part of the knowledge discovery process that we can analyze on an enormous set of data and obtain hidden useful knowledge. It is also the analysis step of the *Knowledge Discovery in Databases* process, or *KDD* in short.

DM is useful in commercial sectors such as financial market analysis and financial prediction, weather forecasting, medicine, transportation, health care, and insurance.

Figure 4.1 shows the famous *data to wisdom hierarchy* that can be found in many IT or data science-related books to conceptualize how humans obtain wisdom (intelligence) from raw data we encounter. The entire process of extracting *data* into *information, information into knowledge,* and *knowledge into wisdom* is identical to data mining tasks and objectives. In other words, DM is the process and technology to extract (so-called *data-mine*) useful knowledge and even intelligence from all possible raw data of different data sources, media, and domains.

Fig. 4.1 Data to wisdom hierarchy in data mining

4.2   Why Data Mining Becomes so Important?

Data mining has practical applications for a range of common business problems. Organizations become efficient to complete tasks by using various data mining techniques. One example could be preprocessing a set of data that requires human intervention at a later stage. Tasks that required lots of user input before can now be automated to some degree. The saved resources can be used for other means.

Beyond task automation, data mining can also be used to analyze large quantities of complex data on predictions as data analysis is an essential task for many businesses. For example, companies analyze sales data to identify profit-making or money-losing segments.

Besides, data mining can allow real-time analysis for complex data and be applied to mission-critical systems potentially. It is an explanatory research and development topic in various disciplines ranging from modern medication to space exploration. Currently, machine learning has a lot of limitations and is far from replacing human requirements. However, the constant evolution of data mining technology can offer solutions to difficult problems that take up too many resources ever considered (Fig. 4.2).

4.3   Knowledge Discovery Process (KDP)

Knowledge Discovery in Databases (KDD) is the process of discovering useful knowledge from data collection, developed by data scientist Dr. Gregory Piatetsky-Shapiro in 1989. This widely used data mining technique is a

Fig. 4.2 Data mining and data analysis

process that includes data preparation and selection, data cleansing, incorporating prior knowledge on datasets, and interpreting accurate solutions from the observed results.

KDD includes the tools and theories to help us extract useful and previously unknown information (i.e. knowledge) from large collections of digitized data (Piatetsky-Shapiro and Frawley 1991).

As shown in Fig. 4.3, a complete KDD process consists of.

1. *Selection*—data selection (e.g. from database);
2. *Preprocessing*—data cleansing and preprocessing;
3. *Transformation*—data transformation to better data mining representation;
4. *Data mining*—data mining of useful information from data (or database);
5. *Interpretation and evaluation*—data evaluation and presentation into higher level knowledge and information for decision-making.

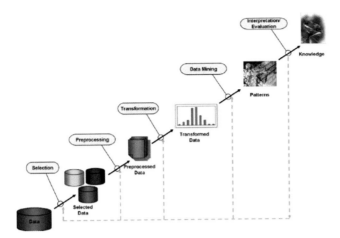

Fig. 4.3 A complete KDD process

KDD consists of several steps and data mining is one of them. Traditionally, data mining and knowledge discovery were performed manually. As time progressed, data amount grew to larger than terabyte size in many systems that could no longer be maintained manually. Moreover, discovering underlying patterns in data is essential for the successful existence of any business. As a result, several software tools were developed to discover hidden data and make assumptions which formed a part of AI. The KDD process reached its peak in the past 10 years. It now houses many different approaches to discovery that includes inductive learning, Bayesian statistics, semantic query optimization, knowledge acquisition for expert systems, and information theory with its ultimate objective to extract high-level knowledge from low-level data.

4.4 Data Preprocessing

Why does data need preprocessing?

Data preprocessing (García et al. 2014) is a data mining technique that involves transforming raw data into an understandable format and is a proven method of resolving problems. We acquire thousands or millions of data (data transactions) from different sources and channels in the real world every day, e.g. sales transactions from a supermarket, patients' information from a hospital, and loan information from a bank. These data are often incomplete, inconsistent, and/or lack certain behaviors/trends, and are likely to contain many errors.

If we take a closer look at the massive data quality, there are multidimensional views that we need to consider:

- Accuracy—data (data transaction) is accurate or not.
- Completeness—data is complete or not.
- Consistency—data itself is consistent with other related data (information).
- Timeliness—data is timely updated. It is particularly important for some time-sensitive data mining problems such as weather or financial predictions.
- Believability—is data credible?
- Interpretability—is data easy to understand?

Basic data preprocessing includes

- Data cleaning,
- Data integration,
- Data transformation, and
- Data reduction (Fig. 4.4).

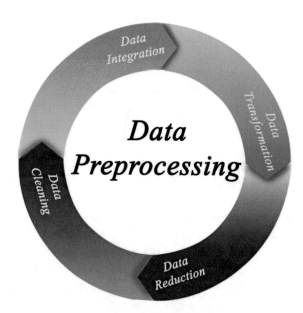

Fig. 4.4 Data preprocessing components (Tuchong 2020a)

4.5   Data Cleaning

There is a large amount of *muddy* and *incorrect* data in the real world, e.g. faulty instrument, human or computer error, and transmission error. *Data cleaning*, also known as *Data Cleansing* (Ilyas and Chu 2019), aims at handling basic problems appearing in the data (raw data) which includes

- *Incomplete data*: lack of attribute values, lack of certain attributes of interest, or contain only aggregate data, e.g. Occupation = "" (missing data);
- *Noisy data*: contain noise, errors, or outliers, e.g. Salary = "−10" (an error);
- *Inconsistent data*: contain discrepancies in codes or names, e.g. Age = "42", Birthday = "03/07/2010" was rating "1, 2, 3", now rating "A, B, C" which is a discrepancy between duplicate records (Fig. 4.5).

How to Handle Missing Data?

- Ignore the tuples: appear when a class label is missing usually (when doing classification)—not effective when % of missing values per attribute varies considerably
- Fill in the missing value manually: tedious + infeasible?
- Fill it in automatically with: a global constant: e.g. using the attribute mean for all samples belonging to the same class or using the most probable value generated from inference-based engines such as Bayesian formula or decision tree.

Fig. 4.5 Data cleaning (Tuchong 2020b)

How to Handle Missing Data?

- *Binning*—First, sort data and partition it into (equal-frequency) bins, then smoothen by the means, median, boundaries, etc.;
- *Regression*—Smoothen by fitting data into regression functions;
- *Clustering*—Detect and remove outliers;
- *Semi-supervised methods*—Combine computer and human inspection. Detect suspicious values and check by human (e.g. deal with possible outliers).

How to Handle Data Inconsistency?

- Detect data discrepancy: Use metadata (e.g. domain, range, dependency, and distribution);
- Check field overload;
- Check uniqueness rule, consecutive rule, and null rule;
- Use commercial tools to perform data scrubbing;
- Use simple domain knowledge (e.g. postal code and spell-check) to detect errors and make corrections;
- Audit data by data analysis to discover rules, relationship and detect violators (e.g. correlation and clustering to find outliers).

4.6　Data Integration

Data integration (Doan et al. 2012) involves combining data from multiple sources into a coherent store for data mining. Data integration allows different data types (i.e. data transactions inside the database, document, and tables) to be merged by users, organizations, and applications to use for personal or business processes and/or functions (Fig. 4.6).

Why Do We Need Data Integration?

- Data is inconsistent and redundant across different data sources (databases), e.g. same person has different name formats across different databases.
- Data is dispersed into different formats and data sources (e.g. texts, audios, videos, figures and graphical formats, or even human dialogues and musical notes).

Fig. 4.6 Data integration from various resources

- Data is dispersed across different databases, OS platforms, and locations (PC server, mobile, cloud, etc.)

Common Methods of Data Integration

- *Data consolidation* is to combine data together from several separate systems physically and create a consolidated data version into one data store. The goal of data consolidation is to reduce the number of data storage locations.
- *Data propagation* is the use of applications to copy data from one location to another. It is event-driven and can be done synchronously or asynchronously. Most synchronous data propagation supports a two-way data exchange between the source and the target.
- *Data virtualization* is to use an interface to provide a near real-time, unified data view from disparate sources with different data models. Data can be viewed in one location, but not stored in that single location. Data virtualization retrieves and interprets data, does not require uniform formatting or a single point of access.
- *Data federation* is technically a form of data virtualization. It uses a virtual database and creates a common data model for heterogeneous data from different systems. Data is brought together and viewed on a single point of access.
- *Data warehousing* is the use of data warehouses that are storage repositories for data. It also involves data cleansing, reformatting, and storage which is basically data integration.

Advantages of Data Integration

- Consolidates data into a more informative format;
- Improves the ease of data/information access and data mining in terms of efficiency and accuracy;
- Reduces access time on different databases and resources during the data mining process;
- Resolves inconsistent and redundant data during data mining;
- Transforms and unifies data into easier access data mining conditions;
- Improves data visualization, user views, and experiences;
- Prepares big data for data mining and data analytical processes.

4.7   Data Transformation

How do we *see* data to understand the information?

Let us have a look at the financial market and we will see how important data transformation is. In short, *Data transformation* (Tan et al. 2018) is the process of changing the format, structure, or values of the data from one form to another in order to (1) make it more easy to understand (for both human and machine) and extract useful information and knowledge; (2) get rid of unrelated and incorrect information and focus on vital and useful information; and (3) transform data that appeared to be completely random or unrelated into useful information and knowledge that can be extracted using data mining technique(s) in high-level perspective.

Figure 4.7a, b shows two presentations of financial data such as *Dow Jones Indices (DJI)* in raw data format and graphical format (financial chart). The truth is, from a financial expert perspective, once financial data is transformed into a financial chart, they can identify many critical financial patterns (such as *Head-and-Shoulder pattern*) and trends based on the so-called technical analysis and chart analysis techniques (Murphy 1999; Bulkowski 2007).

Major data transformation methods include

- Smoothing—focus to remove noise from data. Techniques include data filtering, binning, regression, and clustering.
- Aggregation—focus to summarize data operations, e.g. the daily transactions of a supermarket summarized into weekly and monthly sales to reflect the big picture of sales performance.
- Discretization.
- Normalization.
- Feature extraction.

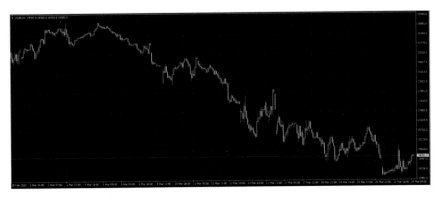

Fig. 4.7 a Dow Jones Index (DJI) in raw data format. **b** Dow Jones Index (DJI) in a financial chart (pattern) format

4.8 Data Discretization

Data discretization (Wohlmuth 2001) is the process of transferring continuous functions, models, variables, and equations into discrete counterparts.

This process is usually carried out as an initial step toward making them suitable for numerical evaluation and implementation on digital computers.

In terms of data preprocessing, a good data discretization method can not only decrease raw data amount to process data mining effectively, but can also reorder (reformat) raw data so that it can reflect more important information and features for data mining.

Basic data discretization methods include.

(1) *Categorization technique*

This technique can be used if the source data can be grouped and categorized into some meaningful classes and categories. For example, the raw data of age groups are usually discretized into interval labels: 9–12 children; 12–19 Youth; 19–30 Young Adults; 30–45 Adults; 45–65 Middle Age; >65 Aged as in Fig. 4.8a.

(2) *Binning technique*

Data binning, also known as *data bucketing,* is a data preprocessing method used to minimize the effects of small observation errors. The original data values are divided into small intervals known as *bins (or buckets)* and then they are replaced by a general value calculated for that specific bin. For example, for the ease of organizing data of a group of products sold in a store, all products are discretized into 11 bins with prices ranging from less than 60 to over 105 as shown in Fig. 4.8b. By

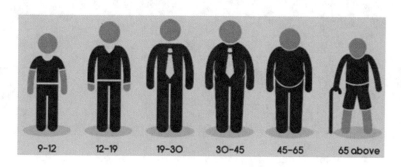

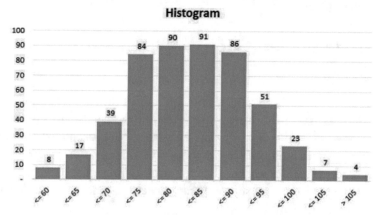

Fig. 4.8 **a** Data discretization with categorization e.g. age group. **b** Data discretization with binning technique, e.g. histogram of data binning

using such discretization method, data in each bin contain sufficient data amount for the data mining process.

Also, the labels used in the discretization method can be organized into higher level concepts, resulting in a concept hierarchy for the numeric attribute. A good discretization method can not only simplify the source data volume, more importantly, it can reorganize raw data to minimize bias by data (data bins) with very few frequencies so that data mining can focus the mining on more representative data (data bins).

4.9   Data Normalization

Data normalization is the process of organizing data (say within a database) to reduce redundancy and improve data integrity. In terms of data transformation, data normalization (Blokdyk 2020) has an additional meaning to transform raw data so that data of different attributes provide an equal weighting and will not be biased by actual data quality.

Let's use a realistic case to demonstrate how it works.

We studied *Artificial Neural Networks (ANN)* (Fausett 1993) which can be used for different AI applications, including weather forecasting. Figure 4.9 illustrates a 7-day weather forecast realistic case using ANN (Taylor and Buizza 2002). ANN required historical time-series weather elements as input that include Air temperature (Temp), Mean-sea-level-pressure (MSLP), Relative humidity (RH), Wind direction (WD), and Wind Speed (WS). As one can see, the number range of different weather elements is completely different. For example, Air Temperature (T) normally ranged between 5 and

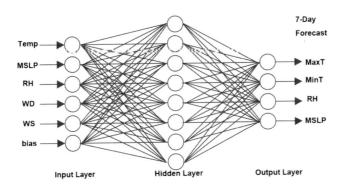

Fig. 4.9   ANN for 7-day weather forecast

35 (°C), Mean-sea-level-pressure is around 9000–12000 (P), and Relative humidity is ranged around 30–98 (%). If we simply feed in all these raw data into ANN, the element with higher quality will usually bias the ANN and result in poor forecasting performance. This is where normalization comes to the picture. Instead of using the actual quality of input data, we normalize all these input elements into numbers ranged between 0 and 1 (or between −1 and +1) so that all input elements will be equally ranged and weighted in the network training and forecast.

Common data normalization methods include

- *Min–max normalization*—the most frequent and simple method used by exercising linear normalization between maximum and minimum values of the data range.
- *Z-score normalization*—this method is commonly used if data distribution falls into a typical normal distribution. In this case, data normalization is evaluated by using the mean (μ) and standard deviation (σ) of the data.
- *Normalization by nonlinear scaling*—the first two methods are functional when data elements are either evenly or normally distributed. But, how about the data that are nonlinearly distributed? In this case, we can exercise a nonlinear scaling function such as an exponential function to transform the data from $[-\infty, +\infty]$ to $[0, 1]$.

4.10 Feature Extraction

Before we study what is *Feature Extraction* (Liu and Motoda 1998), let's begin with a daily scenario on human face identification. Some of us may have bumped into someone in the street who looked familiar to a person we knew. Because of the wrong identification, we apologized with probably an explanation to the person involved who resembled our friend's facial features. We all learnt that human faces consist of important *landmarks* and *features* such as eyes, ears, nose, lips, and cheekbone shape. Throughout human evolution, we have been trained and are capable to extract important facial landmarks and features subconsciously to identify human faces. By adopting this concept, feature extraction aims at the extraction of important features from raw data, particularly useful for complex data such as image patterns.

Common feature extraction methods include

(1) *Landmark feature extraction method* (Lee 2003)—used for data that is able to identify significant and distinct landmarks. They are commonly

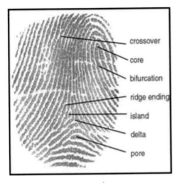

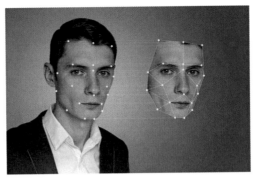

(a) Fingerprint landmarks   (b) Facial landmarks (Tuchong 2020c)

Fig. 4.10 Feature extraction for fingerprint and human face

found in visual patterns and image processing problems such as fingerprints and human face as in Fig. 4.10a, b to show distinct fingerprints and human face landmarks for feature extraction. Feature extraction methods include *wavelet extraction and Principal Component Analysis (PCA)*.

(2) *Feature filtering method* (Liu and Motoda 1998)—used for data that is unable to identify significant and distinct landmarks. In this case, some filtering algorithms would be applied to filter noisy data or extract significant data from the overall data source. These methods are useful on various data sources including image patterns, soundtracks and music, and even massive DNA sequences. The frequently used filtering methods include *Fast Fourier Transform (FFT)* in signal processing, *Gabor filter,* and *Kalman filter*.

(3) *Domain-specific feature extraction and data transformation methods*

The above two feature extraction methods are generic to any data and problem domains. Domain-specific feature extraction and data transformation methods are tailored for specific data mining problems, e.g. lengthy historical and engineering data such as finance or weather analysis and forecasting (Blum 2019).

Here, we use weather forecasting again as an example. Meteorologists transform the surface weather chart into two useful charts: pressure chart of contour line and streamline construction for wind flow instead of data mining weather observation data such as MSLP, temperature, wind speed, and direction shown in Figs. 4.11 and 4.12.

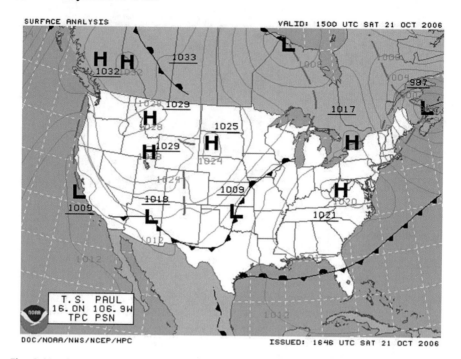

Fig. 4.11 Contour map construction for MSLP in weather chart (Public domain) (Ack: National Oceanic and Atmospheric Administration, NOAA 2020)

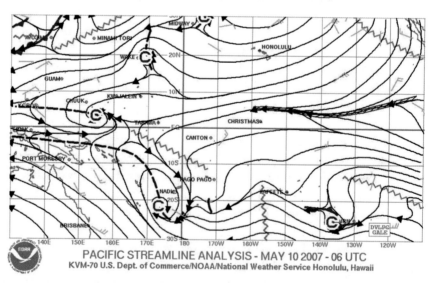

Fig. 4.12 Streamline construction in weather chart (Public domain) (Ack: National Oceanic and Atmospheric Administration, NOAA 2020)

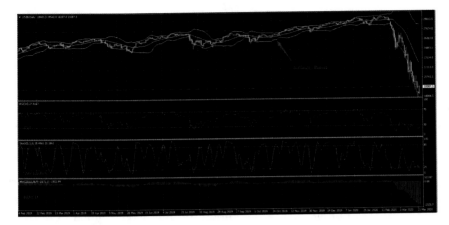

Fig. 4.13 Financial indicators and oscillators

In fact, trained meteorologists can identify high-, low-pressure centers and overall current weather situation by simply observing the streamlines and pressure contour charts.

Financial analysis is another classic example. Technical analysts already constructed a useful financial time-series data transformation method called technical indicators and oscillators frequently used in all financial markets since the 1970s. Typical financial indicators include *Moving Average (MA), Relative Strength Index (RSI), Bollinger Bands (BB), Stochastic oscillators, and MACD indicator.* Figure 4.13 shows a DJI market pattern chart with 4 technical indicators: Bollinger Bands, RSI, Stochastics, and MACD. A trained technical analyst can already gain a big picture and trend (bull or bear) of the current market by observing these financial patterns and technical indicators. These financial indicators and oscillators also provide excellent data mining information (Lee 2020).

4.11 Data Reduction

What is Data Reduction?

A database or data warehouse may store terabytes of data and require a lengthy process to analyze data and mining normally. *Data Reduction* (Bevington and Robinson 2002) techniques can be applied to obtain a reduced dataset representation in a compact volume but contain all critical information.

Common Data Reduction Methods:

(1) *Data compression*—It is the process to reduce the data size by using different encoding mechanisms. It can be divided into two types based on the compression techniques—lossy versus lossless data compression. The main difference between the two compression techniques is that the lossy compression technique cannot restore the data from its original form after decompression, whereas lossless compression can allow data to be restored and rebuilt from its original form after decompression. Typical examples can be found in sound, image, and video compression. *Discrete Wavelet Transform (DWT)* technique and *Principal Component Analysis (PCA)* are examples of these compression methods. A good example of lossy image data compression is the JPEG image format. A common lossy image compression format reduced the image size effectively but preserved all important features from the original image. Figure 4.14 shows a typical example of lossless versus lossy compression of the same landscape image. It is in fact difficult to identify the difference from a lossy image compression format such as JPEG unless we enlarge the image a great deal to compare with the original or lossless image.

(2) *Linear and nonlinear regression*—*Regression method* is a classical statistical method for the generalization of a set of data and observations using either linear or nonlinear functions (or lines in 2D space). Figure 4.15 shows how a set of data (observations) can be represented by a straight

Fig. 4.14 Lossy versus lossless compression of image

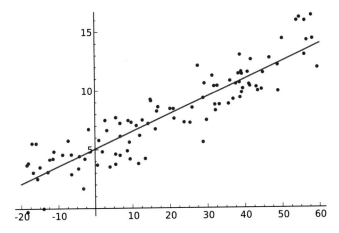

Fig. 4.15 Linear regression on data reduction

line using the linear regression technique. In terms of data reduction, using the regression line to represent the dataset can reduce data size and data mining speed effectively rather than using all observations for data mining. Regression techniques are also frequently used as a data mining tool and method at simple data mining problems.

(3) *Histograms and clustering techniques*—As mentioned in the previous section, histogram provides an easy to use and effective solution for *Data Discretization*. It also provides an excellent task on *Data Reduction*. For instance, to analyze *Daily Returns (r)* of Dow Jones Index (DJI) for the past 2000 trading days, we consolidate these 2000 days' data into 100 intervals between 0.975 and 1.023 as shown in Fig. 4.16. By analyzing and data mining these 100 return intervals, the data size is effectively reduced but the preserved information is contained in the data sources. Also, further data reduction can also be done for this histogram by plotting the nonlinear regression curve over the histogram, shown as a blue regression curve in the figure. In other words, different data reduction methods can be combined to perform preferable data reduction outcomes.

(4) *Data cube aggregation technique*—First of all, what is a *Data Cube*? A data cube is a type of multidimensional matrix that allows users explore and analyze a data collection from many different perspectives, usually consisting of three factors (dimensions) at a time. For instance, there are over 1000 workers and staff in a toy factory. We are assigned to conduct an overall evaluation. How can we do it? We might think to evaluate staff in 3 aspects: *Demographic Data, Organizational Process Data,* and *Predictive Attitudinal Data* and use the data cube method for representation as

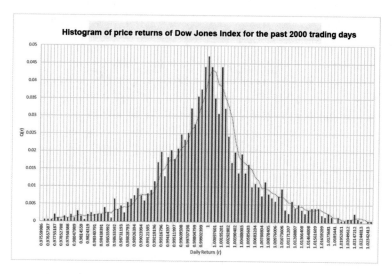

Fig. 4.16 Histogram technique on data reduction

shown in Fig. 4.17. Thus, by using this 3D data cube, the data can be aggregated so that resulting data summarize the evaluation results from different perspectives and interests. Also, the resulting dataset is compact in volume and without loss of information necessary for data analysis.

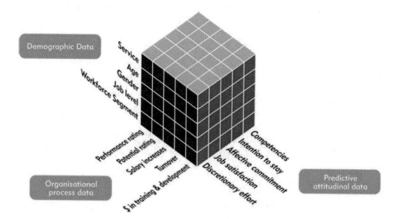

Fig. 4.17 Data cube aggregation on data reduction

4.12 Dimension Reduction

There are often impeding factors to extract data mining sequel in many data mining problems. These factors are intrinsically variables called *features*. The higher the feature number, the harder it gets to visualize and work on the training set. Most of these features are generally correlated and hence redundant. This is where dimensionality reduction algorithms come into the picture.

Dimensionality reduction (Sun et al. 2016) is the process of reducing the random number variables under consideration, by obtaining a set of principal variables. It can be divided into feature selection and feature extraction.

The frequently used dimension reduction methods include wavelet transforms and *principal components analysis (PCA)* which transform or reduce the original data into a compact space. Attribute subset selection is a method of dimensionality reduction to detect and remove irrelevant, weakly relevant, or redundant attributes or dimensions.

A typical example of dimensionality reduction can be found in a simple e-mail classification problem where we need to classify whether the e-mail is spam or not. This involved large numbers of features such as e-mail having a generic title, its contents, and its template. However, some of these features may overlap.

A 3D classification problem can be difficult to visualize, whereas a 2D one can be mapped to a simple 2-dimensional space, and a 1D problem to a simple line. Figure 4.18 illustrates this concept where a 3D feature space

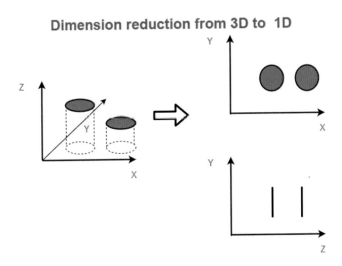

Fig. 4.18 Illustration of the dimension reduction scheme

is split into two 1D feature spaces and later, if found to be correlated, the number of features can be further reduced.

4.13 Classical Methods of Data Mining

There are many methods used for Data Mining (Tan et al. 2018) but the crucial step is to select the appropriate method according to the business or the problem statement. Figure 4.19 illustrates the basic data mining methods with four major tasks:

- Classification,
- Clustering,
- Regression, and
- Association.

Classification takes present information and merges it into defined groupings. *Decision tree method* is a frequently used method.

Clustering removes the defined groupings and allows data to classify itself with similar items. *K-means method* is a frequently used method.

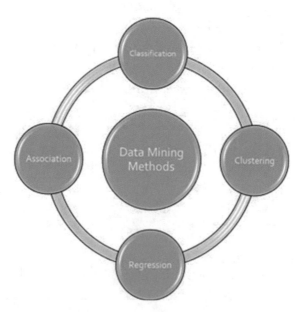

Fig. 4.19 Major classical data mining methods

Regression focuses on the function of information, modeling the data on the concept. *Linear regression method* is a frequently used method.

The final data mining method, association, attempts to find relationships between various data feeds. Association Rules method is a frequently used method. These data mining methods help to analyze market trends, predict future courses, make decisions, and increase company revenue accordingly.

4.14 Classification Using Decision Tree

What is a Decision Tree?

Decision tree (Tan et al. 2018; Rokach 2014) is one of the most powerful and popular tools for classification and prediction. A decision tree is a flowchart-like tree structure, where each node denotes a specific attribute, each branch of the decision tree represents a possible decision, outcome, or reaction. The farthest branches on the tree represent the end results. It is the most frequent data mining method used to clarify and find an answer to a complex problem such as banking, finance, investment, and business.

An organization may deploy decision trees as a kind of decision support system. The structured model allows the chart reader to see how and why one choice may lead to the next, with the use of branches to indicate mutually exclusive options. In other words, the structure allows users to take a problem with multiple possible solutions and display those solutions in a simple, easy-to-understand format and indicate the relationship between different events or decisions.

In a decision tree, each end node has an assigned weight, frequency, or probability. Users look at each terminal outcome for weighting and probability assessment. The tree can span its length as needed until it comes to a proper conclusion.

How to Construct a Decision Tree?

There are many methods to create a decision tree, the most often used method being *Frequent Pattern Classification Technique* based on database records and attributes in the database for decision tree generation. This method is frequently used in major organizations such as banks, insurance companies, and financial institutions.

In this section, we use a realistic case of loan approval from a bank to create a simple data mining and decision support system based on previous loan approval transactions available in the past 1 year.

Figure 4.20 shows the loan approval transaction for 30 cases with four attributes:

Rec_ID	Age	Employed	Own_house	Credit_Rating	Loan Approved
1	Young Adult	Yes	Yes	Fair	No
2	Young Adult	Yes	Yes	Good	Yes
3	Young Adult	Yes	No	Good	Yes
4	Young Adult	Yes	No	Excellent	Yes
5	Young Adult	No	No	Fair	No
6	Young Adult	No	Yes	Good	No
7	Young Adult	No	No	Good	Yes
8	Young Adult	No	No	Fair	No
9	Young Adult	No	No	Excellent	Yes
10	Middle Age	Yes	Yes	Excellent	Yes
11	Middle Age	Yes	No	Good	Yes
12	Middle Age	Yes	No	Excellent	Yes
13	Middle Age	Yes	Yes	Good	Yes
14	Middle Age	Yes	Yes	Excellent	Yes
15	Middle Age	Yes	Yes	Good	Yes
16	Middle Age	Yes	Yes	Fair	No
17	Middle Age	Yes	Yes	Good	Yes
18	Middle Age	No	Yes	Good	No
19	Middle Age	No	No	Fair	No
20	Middle Age	No	Yes	Fair	No
21	Aged	Yes	Yes	Excellent	Yes
22	Aged	Yes	Yes	Good	Yes
23	Aged	No	Yes	Fair	No
24	Aged	No	Yes	Good	Yes
25	Aged	No	Yes	Good	Yes
26	Aged	No	No	Excellent	Yes
27	Aged	No	No	Fair	No
28	Aged	No	Yes	Fair	No
29	Aged	No	Yes	Good	Yes
30	Aged	No	No	Fair	No

Fig. 4.20 Example of a data record in credit rating

- Age: *Young Adult, Middle Age, or Aged;*
- Employed: *Yes* or *No;*
- Own house: *Yes* or *No;*
- Credit rating: *Fair, Good, or Excellent.*

Figure 4.21 shows a 2-level decision tree on Age (level 1) and *Employed/Own House/Credit Rating* (level 2). The final loan approval decision is obtained by the following equation:

(No. of approval/disapproval cases)/(total no. of cases that meets the requirement).

By using the 2-level decision tree, we can data-mine a query as follows:

Case 1: Young Adult + Employed → Approval rate 3/4= 75%

Case 2: Middle Age + Not own house → Approval rate 5/8 = 62.5%

Case 3: Aged + Good Credit Rating → Approval rate 4/4 = 100%

So, how about a more complex case with 3-level condition?

Figure 4.22 shows a 3-level decision tree with attributes Middle Age (level 1), Employed or not (level 2), and Credit Rating (level 3).

From this 3-level decision tree, we can support the following queries:

Case 4: Middle Age + Employed + Fair credit rating → Disapproval rate 3/3 = 100%

Case 5: Middle Age + Employed + Good credit rating → Approval rate 4/5 = 80%

Case 6: Middle Age + Employed + Excellent credit rating → Approval rate 3/3 = 100%

In this decision tree, we can also drive another important case:

Case 7: Middle Age + Not Employed → Disapproval rate 3/3 = 100%

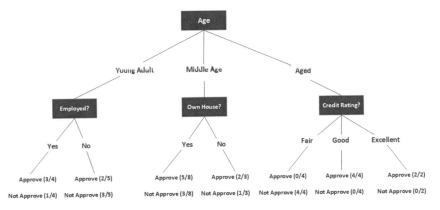

Fig. 4.21 2-level decision tree in credit rating

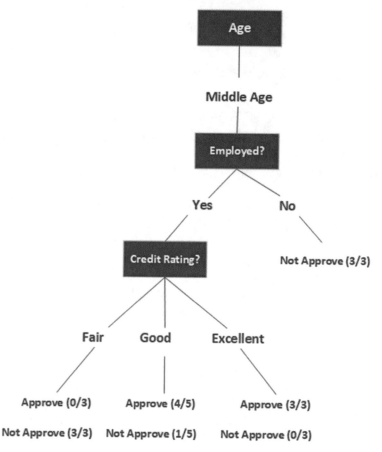

Fig. 4.22 3-level decision tree in credit rating

which means that the client is middle age, unemployed, the decision is disapproved regardless of credit rating.

4.15 Clustering Using KNN Method

What is Clustering?

There is another kind of data mining method, apart from decision tree, we use frequently in daily activities—*Classification and Clustering* (Aggarwal and Reddy 2013).

Let us begin with our typical daily life cycle on a normal working day: We wake up, go to a coffee shop for breakfast, our first classification and clustering task is to select our favorite bread or muffin. How do we select? First, we cluster unconsciously our favorite breakfast food (e.g. muffin) out of other food varieties. Second, we classify consciously different muffins' outlook to select the best one. Afterwards, we take a bus or train to work. We then classify consciously buses or train compartments into different classes or clusters to select the least pressing one to climb aboard. When we arrive and enter the office building, we again classify or cluster consciously or even unconsciously different queues and select the shortest one to take the elevator. We apply the same technique at our work throughout the day.

For example, no matter what kind of work we do, we would probably deal with a typical problem: What kind of task(s) to begin with? What is the priority of the tasks? Different individuals use different methods but naturally, most of them are related to classification and clustering in terms of urgency, importance, time-dependency, etc. As we can see, each of them is a typical kind of classification or clustering. We classify these tasks frequently according to different attributes and make the best decision simultaneously. Figure 4.23 shows a typical example of original data versus clustered data into 3 classes (groups).

K-Nearest Neighbor (KNN) Technique

In data mining, one of the most popular and important classification and clustering methods is the *K-Nearest Neighbor (KNN)* technique. The concept of the KNN technique is simple: an object or instance (o) belongs to one

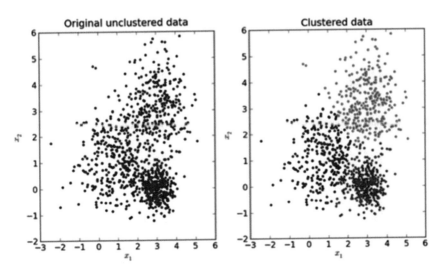

Fig. 4.23 Original data versus Clustered data

specific class (c). If we try to measure the distance (d) between this object (o) with all objects of different classes and arrange them according to distances, for a KNN classification strategy, we only consider the first K object with the shortest distance and count the number of objects within each class. The class (o) with the maximum number will be the one this object (o) belongs to.

Here, we use a realistic example of students' classification/clustering of a university to illustrate how the KNN method works. To simplify the problem, we assume the student population belongs to 3 different faculties/groups Arts, Management, and Science with three different object shapes and colors as shown in Fig. 4.24.

To do classification, we must have some *attributes* for consideration. Here, we use 2 attributes for illustration purpose, *Logical Thinking* and *Artistic Interest* so that it can be shown in a 2D figure. There can normally be 3 or more consideration attributes simultaneously in a real-world situation.

The classification/clustering problem is: *How can we classify and cluster new student A in that population, provided that we have his scores for Logical Thinking and Artistic Interest using the KNN method?*

Case 1—Using KNN for K= 5:

KNN-5 is shown in Fig. 4.25 where the unknown student is marked with letter "A". We have 2 methods to perform KNN clustering for our problem

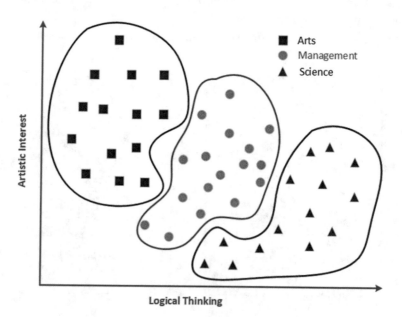

Fig. 4.24 Clustering of students into 3 groups

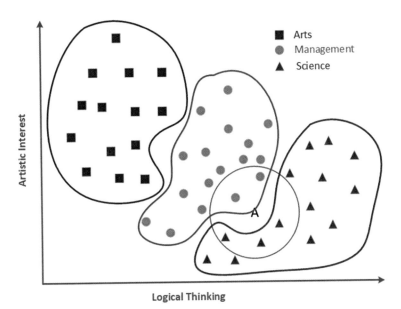

Fig. 4.25 Case 1—Student A with KNN-5

with 2 attributes: (1) Use the revealed algorithm and calculate all distances between A and all student objects in the figure, sort them according to distance and count first 5 student objects with the shortest distance, and identify which faculty has the maximum student numbers; or (2) simply draw a circle with student A as the center that contains exactly 5 students objects, and count which color objects are with the maximum count. As shown, a circle created includes exactly 5 student objects; the number of counts for each faculty is *Arts (0), Management (2), and Science (3)*. Therefore, by using KNN = 5, the unknown student should belong to the Science faculty.

Case 2—Using KNN for K = 13:

One might ask: Is it always the same result for different K number? The answer is a definite *no*. Figure 4.26 illustrates the clustering result when KNN = 13. Using the fast circling method, we can see the result: *Arts (0), Management (7), and Science (6)*. By using KNN = 13, the unknown student should belong to the Management faculty.

Why? It all depends on data, priori knowledge, and data distribution. In other words, the choice of K number is critical for data mining. It relies on data analyst/data scientist's experience to perform the data mining task.

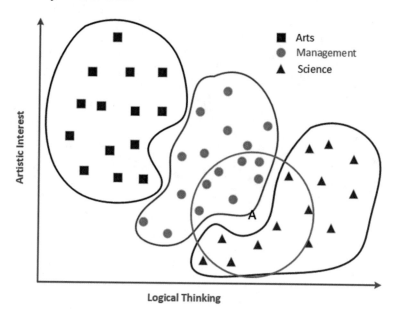

Fig. 4.26 Case 2—Student A with KNN-13

One might ask: *How can we create 3 clusters at the very beginning?* It would be a *chicken-and-egg problem.* If we do not have the initial data, there wouldn't be any clusters, and without the clusters there wouldn't be any classification. The truth is, for any classification problem, we must have 2 basic elements: (1) Some basic objects that have predefined classes to begin with and there will be no cluster; (2) Some basic (at least 2) attributes for evaluation and there will be no way to do distance measurement if otherwise. In fact, once we have both (1) and (2), all KNN classification procedures for all unknown objects will be done and will enlarge the clustering class of the population.

KNN is a frequently used method in many real-world data mining problems on population classification, e.g. relationship between occupation and education levels, demographic analysis, voting preference distribution and clustering, and even to a new eatery location.

4.16 Regression Method

What is Regression?

Regression (Matloff 2017) is a useful statistical-based data mining tool to estimate the relationships between a *dependent variable* (often called the *outcome variable*) and one or more *independent variables* (*features or attributes*).

In terms of data mining and knowledge discovery, regression is used across multiple industries for business and marketing planning, financial forecasting, environmental modeling, and trend analysis.

Classification Versus Regression

Regression and *classification* are both data mining techniques used to solve similar problems, but they are often confused. Although both can be used on prediction analysis, regression is used to predict a numeric or continuous value while classification assigns data into discrete categories.

For example, regression would be used to predict a house price based on the location, size, and environmental factors which are continuous values. On the other hand, classification would be applied for users to choose different house types such as apartment, semi-detached, detached house or a bungalow based on price, location, and environmental factors for which 4 different house types are discrete classes and categories.

Linear Regression

The most basic and simple type of regression is called *Linear Regression*. Figure 4.27 illustrates a case of linear regression for a set of experimental observations with two variables: the *independent variable* x (attribute) and the *dependent variable* (outcome).

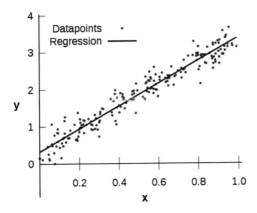

Fig. 4.27 Linear Regression for a set of experimental observations

Linear regression uses the mathematical formula of a straight line ($y = mx + b$). In plain terms, given a graph with a Y- and an X-axis, the relationship between X and Y is a straight line with a few outliers.

The logic behind this is that: *Given a population of observations, the straight line with a few outliers provides the best fit and relationship estimation between the attribute and outcome of the data mining problem.*

Figure 4.28 illustrates the linear regression between the *Total Bill* versus *Tip of the bill* of a restaurant as a real-world problem. We might assume that, given an increase of *Total Bill*, there would be an increase in the *Tip* amount in a linear manner. To visualize this, consider a graph in which the Y-axis tracks the increase of *Total Bill*, and the X-axis tracks the *Tip*. As Y value increases, X value would increase at the same rate, making the relationship between them a straight line.

This also illustrates how a linear regression line is formed. As shown, for each observation data, the vertical distance between the observation point and the straight line is known as *deviation*. So, the regression line is formed by finding a straight line such that the sum of deviations from all observation data is the minimum with the method called *least square approach.*

Although the calculation is quite tedious, thanks to the popularity of linear regression and computer technology, linear regression is almost a standard tool embedded in many applications ranging from Excel in MS Office to data mining and statistical tools such as R, Python, and MATLAB.

Nearly all real-world regression models involve multiple attributes, and basic linear regression is often phrased in terms of a *multiple regression model* (or *multivariable linear regression*).

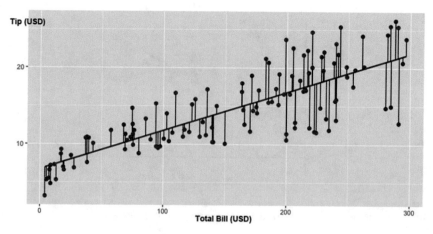

Fig. 4.28 Linear regression of *total bill versus tip*

Nonlinear Regression

Nonlinear regression is a form of regression analysis in which data is fit to a model and then expressed as a mathematical function. Simple linear regression relates two variables (X and Y) with a straight line ($y = mx + b$), while nonlinear regression must generate a line (typically a curve) as if every value of Y is a random variable. Figure 4.29 illustrates a typical nonlinear regressive curve.

An example of how nonlinear regression can be used is to predict population growth over time. A scatterplot of changing population data over time showed that there seemed to be a relationship between time and population growth, but that is a nonlinear relationship that is required to use a nonlinear regression model. A logistic population growth model provided population estimates for periods that were not measured, and future population growth predictions. Opposite the nonlinear population growth curve is the famous *decay curve*, with the name coming from the radiative decay of radiative elements such as Uranium 235. Figure 4.30 illustrates a typical nonlinear radiative decay curve. In fact, the decay curve is commonly found in many real-world situations such as the market economy, finance, biology, and telecommunication.

Besides population growth and decay curve, another type of nonlinear regression curve commonly found in many data mining problems is *logistic curve*, also known as *sigmoid curve* with its distinctive S-shape. In fact, the sigmoid growth curve is believed to be a natural phenomenon commonly

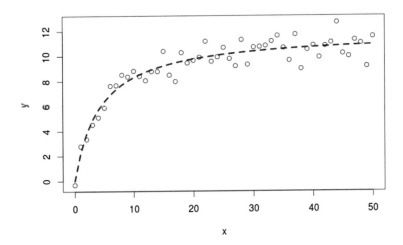

Fig. 4.29 Nonlinear regression population curve

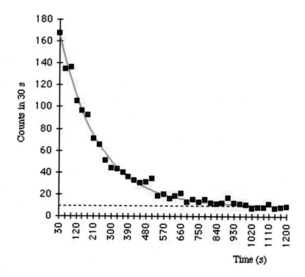

Fig. 4.30 Nonlinear regression radiative decay curve

found in many natural science, biology, and medical tests such as drug response experiments. Figure 4.31 illustrates nonlinear sigmoid curves of two vaccine (*Bupivacaine and Ropivacaine*) dosages versus responses of patients' pain relief scores.

As shown, both vaccines demonstrated typical logistic growth with exceptionally low responses until they reached the first threshold dosage, then the responses grew exponentially until they reached the second threshold dosage. After that, even higher dosages were unable to produce any significant

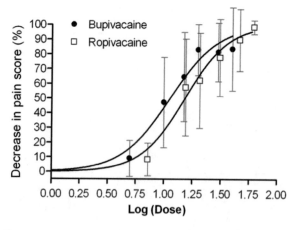

Fig. 4.31 Nonlinear regression sigmoid curve on vaccine responses

improvement. In other words, an effective vaccine can be clearly identified by using a nonlinear regression technique. More importantly, we could also predict the degree of effectiveness with different dosages and select the optimum dosage for patients.

4.17 Association Rule Method

What is Association Rule?

In many real-world scenarios, we always want to data-mine patterns, either consciously or subconsciously. These *patterns* can be anything related to two or more attributes, facts, objects, outcomes, or even thoughts in general.

For example, we bring an umbrella when the sky is cloudy and windy. It is natural and common sense but it is a typical kind of *Association Rule* (Tan et al. 2018; Adamo 2012) in the form: X → Y where X is the *cause* (or *condition*) and Y is the *effect* (or *outcome*). In general, X can be a single condition or multiple conditions, and Y can be anything. It can be an action, decision-making, event(s) outcome, or even a condition for another association rule. In terms of data mining, we always named condition elements as items, and a group of conditions as itemsets. So, in our example, the association rule will be {cloudy, windy} → {bring umbrella}.

In addition to the above simple daily activity example, association rule has more powerful use in countless daily examples. One of them is data mining of *customer habits* in a supermarket.

In this section, we will try to use a simple case of data mining customer purchase transactions to illustrate how association rule pertains to data-mine customer habit.

TID	Transaction List
01	Coke, Nuts, Chips
02	Coke, Coffee, Chips
03	Coke, Chips, Eggs
04	Nuts, Eggs, Milk
05	Nuts, Coffee, Chips, Eggs, Milk

Fig. 4.32 Transaction list of 5 customers in a supermarket

Figure 4.32 shows a transaction list of 5 customers at a supermarket where *TID* is the *Transaction ID*. The list of purchase items included *Chips (Potato chips), Nuts, Coke, Coffee, Eggs, and Milk.*

Before we begin, let us establish some basic definitions using the *Association Rule* method:

- Itemset is the set that contains one or more items, e.g. {Chips}, {Chips, Coke}.
- K-itemset is the itemset that contains k items, e.g. {Chips, Coke, Milk} is a 3-itemset.
- Support(itemset) (also written as "sup(itemset)") is the frequency of an itemset X in transaction dataset,
    e.g. sup(Chips) = 3, sup(Nuts) = 3, sup(Coke) = 4, sup(Chips, Coke) = 3, and sup(Chips, Nuts) = 1.
- Fraction support s{X} is the probability that a transaction contains itemset X,
    e.g. s{Chips} = 3/5 = 60%, s{Coke} = 4/5 = 80%, s{Chips, Coke} = 3/5 = 60%, s{Chips, Milk} = 1/5 = 20%.

In short, the complete Association Rule Mining is a 3-step process:

(1) Identifying patterns (itemsets);
(2) Generating association rules from frequent patterns;
(3) Mining association rules.

What is a Frequent Pattern?

An itemset (or a *pattern*) X is *Frequent* if the support of X is not less than the *minsup* threshold σ.

In a typical data mining case, we will set $\sigma = 50$ (i.e. 50/50) at the beginning. So, in our 5-transaction dataset:

(1) All frequent 1-itemsets are
        s{Chips}: 3/5 (60%), s{Nuts}: 3/5 (60%), s{Coke} 4/5 (80%), s{Eggs}: 3/5 (60%)
        (Note: Milk and Coffee are excluded as their s{Milk} = s{Coffee} = 2/5 = 40% are less than minsup threshold σ.
(2) All frequent 2-itemsets are

s{Chips, Coke}: 3/5 (60%)
(Note, all other s{} of the 2-itemsets are less than σ).

(3) All frequent 3-itemsets (or above)?—None.

How to Generate Association Rules?

To generate association rule, we make sure of the frequent patterns. As mentioned at the beginning of this section, all association rules are in fact causations (cause-and-effect) generated by our previous knowledge or experiences. In terms of dataset data mining, they are simply the "if–then" kind of association between different patterns (or itemsets).

For example, {Coke} → {Chips} means "Customer who buys Coke may likely buy Chips".

So, one might wonder: How strong is this association rule?

To answer the question, we define two thresholds: Support (s) and Confidence (c) as a kind of evaluation of an association rule, like this: X → Y(s, c) where X and Y are patterns (itemsets).

Support (s)—the probability that a transaction contains X ∪ Y, or the number of the transaction contains both X and Y over the total number of transactions in the dataset as illustrated in Fig. 4.33.

For example, s{Coke, Chips} = 3/5 = 60%.

Confidence (c)—the conditional probability that a transaction X also contains Y.

Based on our previous definition of sup{},

$$c = \text{sup}\{X \cup Y\}/\text{sup}\{X\} = s\{X \cup Y\}/s\{X\}.$$

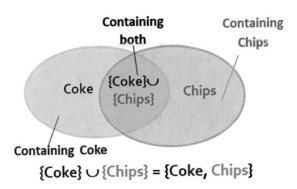

$\{Coke\} \cup \{Chips\} = \{Coke, Chips\}$

Fig. 4.33 Illustration of the union between itemsets Chips and Coke

In our example,

$c = \sup\{Coke, Chips\}/\sup\{Coke\} = 3/4 = 75\%$

How to Perform Association Rule Mining?

The basic steps are straightforward:

(1) Create TWO thresholds: minsup (minimum support) and minconf (minimum confidence).
(2) Generate all possible association rules $X \rightarrow Y$ (s, c) from itemsets and transaction datasets.
(3) Extract association rules $X \rightarrow Y$ (s, c) that meet the minsup and minconf thresholds, such that $s \geq$ minsup and $c \geq$ minconf.

In our example,

Set minsup = 50% (50/50, default case).
All possible Frequent 1-itemset: $s\{Chips\} = 3/5(60\%)$, $\sup\{Nuts\} = 3/5(60\%)$,, $s\{Coke\} = 3/5(80\%)$, $s\{Eggs\} = 3/5(60\%)$.
All possible Frequent 2-itemset: $s\{Chips, Coke\} = 3/5(60\%)$.
Set minconf = 50% (50/50, default case)
All possible transaction rules that meet minconf threshold are
$\{Chips\} \rightarrow \{Coke\}$ with $c = s\{Chips \cup Coke\}/s\{Chips\}$
$= \sup\{Chips \cup Coke\}/s\{Chips\} = 3/3 = 100\%$,
so $\{Chips\} \rightarrow \{Coke\}$ (60%, 100%)
$\{Coke\} \rightarrow \{Chips\}$ with $c = s\{Chips \cup Coke\}/s\{Coke\}$
$= \sup\{Chips \cup Coke\}/s\{Coke\} = 3/4 = 75\%$,
so $\{Coke\} \rightarrow \{Chips\}$ (60%, 75%)

This simple example corresponds to an interesting but important data-mine association rule about customer habit: Customers buying potato chips would most likely buy Coke as well, but NOT the other way round.

How Important Association Rule Data Mining Is?

In fact, most of the *cross-selling strategies* in major departmental stores or online stores use various kinds of data mining strategies to data-mine the association rules between different items to push for overall sales performance. Imagine in a supermarket or online store, there are thousands or up to millions of transactions per minute. If we can data-mine the top 100 association rules with over 85% confidence level and the performance of selling cross products, how much extra selling proposition can we generate?

There is one closing point to indicate that in a real-world situation, we have over millions of items and millions of transactions per day. Then how can we generate all these association rules automatically?

The reason is twofold: (1) After the design of association rules materialized long ago, we already have some excellent algorithms to generate association rules such as Apriori algorithm and FP-growth *(FP means frequent patterns)* algorithm to automate the entire FP mining process; (2) Thanks to the rapid development of data mining technology, we now have powerful and useful tools such as Python and R to perform the association rule mining automatically. All we need is basic concepts of how association rule works, and more importantly, to have a better knowledge of our problem domain. The truth is, even if we have the best data mining tools to establish data-mine association rules, we still need to interpret the rules and determine actions to execute decisions.

4.18 Deep Neural Networks for Data Mining

In Chap. 3, we had learnt different types of AI-based machine learning methods and technologies. Can these methods be applied to data mining too?

The answer is *yes*.

Artificial neural networks (ANN), especially *multilayer neural networks* (Aggarwal 2018; Lee 2006), can definitely be used for data mining ranging from different cluster classifications to time-series weather prediction. Figure 4.34 illustrates the neural architecture of a typical multilayer neural network.

However, due to the complexity and massive data volume in many real-world data mining problems such as real-time financial predictions of worldwide financial markets which involve the handling of numerous financial data across different financial markets, i.e. *forex* (foreign exchange), *commodities* (e.g. gold, silver, crude oil, cotton, and sugar), and *financial indices* (e.g. DJI, HIS, and FT100), certain enhancements of classical *multilayer neural networks* are required where *deep neural networks (DNN)* comes into the picture.

Input Layer   **Hidden Layer**   **Output Layer**

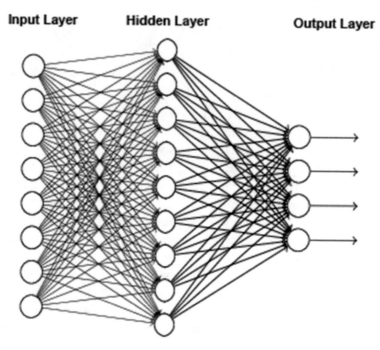

Fig. 4.34 Neural architecture of multilayer neural networks

A *Deep Neural Network (DNN)* is an *artificial neural network (ANN)* with multiple hidden layers between input and output layers. The machine learning performed by DNN is also known as *deep learning*. In deep learning, the number of hidden layers can even reach up to 100 layers. Current research revealed that DNN produces much better results than normal ML networks (Lee 2020).

The main logic behind this is that if a classical neural network contains more hidden layers, technically speaking, it can learn more complex knowledge and accept more massive input data for machine learning and data mining.

So, how does it work?

Figure 4.35 illustrates the network architecture of DNN constructed by the author for daily worldwide financial market prediction in *quantum finance forecast center* (QFFC 2020).

As shown, the *input layer* accepts daily time-series worldwide financial data, together with all the related financial indicators and feeds into an 8-level *bifurcation hidden layer (BHL)* for deep learning to forecast the next-day *open, high, low, and closing prices.*

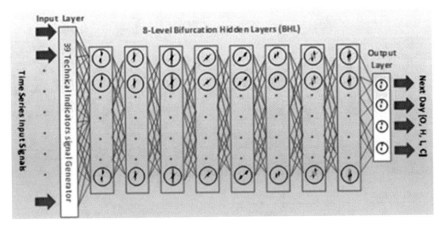

Fig. 4.35 Deep neural network for financial prediction (Lee 2020)

Since Dec 2018, *quantum finance forecast center (QFFC)* applied DNN to predict the next-day forecast for over 120 worldwide financial markets including 9 major cryptocurrencies, 84 forexes, 19 major commodities, and 17 worldwide financial indices across different countries. Figure 4.36 shows the official site and daily financial forecast on Mar 27, 2020.

Besides the financial forecast, DNN can be applied in various complex data mining problems such as

(1) *Financial fraud detection* (Baesens et al. 2016)—Deep learning is being applied successfully to financial fraud detection and anti-money laundering. *DNN-based anti-money laundering detection system* can data-mine relationships and similarities between money transaction flows and detect anomalies or, classify and predict specific patterns. The solution leverages both classification of suspicious transactions, and anomaly transaction flow alert and detection (Fig. 4.37).

(2) *Customer relationship management* (Berry and Linoff 2008)—DNN has been used successfully in the market to approximate the value of possible direct marketing actions, defined in terms of *RFM* variables: *Recency*— How recently do customers purchase? *Frequency*—How often do they purchase? and *Monetary Value*—How much do they spend?

(3) The estimated value function was shown to have a natural interpretation as customer lifetime value and be able to data-mine effectively using DNN technology.

(4) *Drug discovery and toxicology* (Hoffmann et al. 2013)—A large percentage of candidate drugs fail to win regulatory approval. These failures are caused by insufficient efficacy (on-target effect), undesired interactions

Fig. 4.36 Deep neural network for daily worldwide financial prediction in quantum finance forecast center (QFFC.org)

Fig. 4.37 Financial fraud detection

(off-target effects), or unanticipated toxic effects. Research has explored the use of deep learning to predict the biomolecular targets, off-targets, and toxic effects of environmental chemicals in nutrients, household products, and drugs. Current research revealed that DNN such as

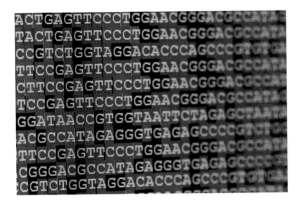

Fig. 4.38 DNA sequencing

AtomNet can be used effectively on structure-based rational drug design, and more importantly used to predict novel candidate biomolecules for target diseases such as Ebola virus and multiple sclerosis.

(5) *DNA sequencing and pattern discovery* (Shui 2016)—DNA sequencing is the process of determining the nucleic acid sequence—the order of nucleotides in DNA. It includes any method or technology that is used to determine the order of the four bases: adenine (A), guanine (G), cytosine (C), and thymine (T) as shown in Fig. 4.38. The advent of rapid DNA sequencing methods has greatly accelerated biological and medical research and discovery. Recently, DNNs were proved to be able to extract useful features from input patterns from DNA automatically, which shed light on DNA sequence classification improvement. They also provide vital help to discover the association between DNA anomalous patterns and certain critical diseases such as various kinds of cancers. DNN can also be applied to other AI problems such as *Computer Vision (CV)* and *Natural Language Processing (NLP)* which will be explored in the following chapters.

4.19 Case Study: Where to Open a New Pizza Shop?

Data mining provides a powerful tool for us to classify, cluster, and even examine customers' habits and sales predictions. In this case study, suppose we are the Data Scientist of a pizza chain store with over 100 pizza shops across the US, and are assigned to provide all the necessary data mining information for management to open a new pizza shop in the US. The objective is to make use of different data mining methods learnt in this chapter such as classification, clustering, decision tree, association rule, and deep neural networks to achieve the following tasks:

(1) Where is the best location to open a new shop?
(2) What kinds of food to be sold in the new shop?
(3) How can we do future 1-year sales projection of the new shop?
(4) How can we improve the sales performance of the new shop by determining the cross-products sale relationship sold in chain shops of the past 2 years?

For example, to select the location, we might consider classification and clustering of demographics across different states in the USA. Figure 4.39 shows a demographic chart of pizza locations across the US in colors, with an illustration of the TOP 10 states in the US.

Fig. 4.39 Demographic chart of pizza shops across different states in the USA

Hints: We might consider the following data mining tasks to solve the problem:

- How to use classification and clustering methods such as KNN to pinpoint the best location?
- How to use decision tree to decide what kinds of food to be sold?
- How to use association rule construction to data-mine customers' eating habits?
- How to use a regression technique to perform sales projection?
- How to use the DNN technique to improve sales performance?

4.20 Conclusion

In this chapter, we study and explore various data mining methods and technologies. They include KNN for clustering, decision tree for decision-making, regression for forecast and projection, and association rule for mining useful patterns. We also review the latest AI technology—deep neural networks (DNNs).

The truth is there are many other useful data mining tools and technologies, such as *Support Vector Machine (SVM)*, *Principle Component Analysis (PCA)*, *Bayesian networks*, and *Genetic Algorithms (GA)*. In fact, the concepts and scope of data mining are enormous to be a self-contented undergraduate course in many universities. The focus of this chapter is to provide readers an overview of concepts and key technologies, more importantly, on how it can be applied to our real-world daily activities.

A practical solution should involve data mining and knowledge discovery from different data sources, data formats, and appearances.

It is evident that data mining not only is one of the most competitive topics for R&D in AI and scientific world, but it also plays a pivotal role such that many major corporations such as banks, insurance companies, and government departments nowadays have designated Data Science divisions (departments) with hundreds of *Data Scientists* to data-mine intensive business knowledge or the so-called *Business Intelligence (BI)* to improve customer service and business development (Fig. 4.40).

Fig. 4.40 Data mining from different perspectives (Tuchong 2020d)

References

Adamo, J. M. (2012). *Data mining for association rules and sequential patterns: Sequential and parallel algorithms.* Springer.

Aggarwal, C. C. (2018). *Neural networks and deep learning: A textbook.* Springer.

Aggarwal, C. C., & Reddy, C. K. (2013). *Data clustering: Algorithms and applications* (Chapman & Hall/CRC Data Mining and Knowledge Discovery Series Book 31, Kindle edition). Chapman and Hall/CRC Press.

Baesens, B. et al. (2016). *Fraud analytics using descriptive, predictive, and social network techniques: A guide to data science for fraud detection* (Wiley and SAS Business Series). Wiley.

Berry, M. J. A., & Linoff, G. S. (2008). *Mastering data mining: The art and science of customer relationship management.* Wiley.

Bevington, P., & Robinson, D. K. (2002). *Data reduction and error analysis for the physical sciences* (3rd ed.). McGraw-Hill Education.

Blokdyk, G. (2020). *Data normalization: A complete guide* (Kindle edition). 5STAR-Cooks.

Blum, A. (2019). *The weather machine: A journey inside the forecast* (Kindle edition). Ecco.

Bulkowski, T. N. (2007). Encyclopedia of chart patterns (Wiley Trading Book 347, Kindle edition). Wiley.

Doan, A. et al. (2012). *Principles of data integration.* Morgan Kaufmann.

Fausett, L. V. (1993). *Fundamentals of neural networks: Architectures, algorithms and applications.* Pearson.

García, G. et al. (2014). *Data preprocessing in data mining* (Intelligent Systems Reference Library Book 72). Springer.

Hoffmann, R. D. et al. (2013). *Data mining in drug discovery* (Methods and Principles in Medicinal Chemistry Book 57). Wiley-VCH.

Ilyas, I. F., & Chu, X. (2019). *Data cleaning.* ACM Books.

Lee, R. S. T. (2020). *Quantum finance: Intelligent forecast and trading systems.* Springer.

Lee, R. S. T. (2006). *Fuzzy-neuro approach to agent applications: From the AI perspective to modern ontology.* New York; Berlin: Springer.

Lee, R. S. T. (2003). *Invariant object recognition based on elastic graph matching: Theory and applications.* IOS Press.

Liu, H., & Motoda, H. (1998). *Feature extraction, construction and selection: A data mining perspective* (The Springer International Series in Engineering and Computer Science Book 453, Kindle edition). Springer.

Matloff, N. (2017). Statistical regression and classification: From linear models to machine learning (Chapman & Hall/CRC Texts in Statistical Science). Chapman and Hall/CRC Press.

Murphy, J. J. (1999). *Technical analysis of the financial markets: A comprehensive guide to trading methods and applications* (New York Institute of Finance, Kindle edition). Prentice Hall Press.

NOAA. (2020). *Official site for National Oceanic and Atmospheric Administration.* Retrieved May 6, 2020, from https://www.noaa.gov.

Piatetsky-Shapiro, G., & Frawley, W. (1991). *Knowledge discovery in databases* (American Association for Artificial Intelligence). AAAI Press.

QFFC. (2020). *Quantum Finance Forecast Center official site.* Retrieved May 6, 2020, from https://qffc.org.

Rokach, L. (2014). *Data mining with decision trees: theory and applications* (Series in Machine Perception and Artificial Intelligence, 2nd ed.). World Scientific Publishing Company.

Shui, Q. Y. (2016). *Big data analysis for bioinformatics and biomedical discoveries* (Chapman & Hall/CRC Mathematical and Computational Biology). Chapman and Hall/CRC Press.

Sun, L. et al. (2016). *Multi-label dimensionality reduction* (Chapman & Hall/CRC Machine Learning & Pattern Recognition). Chapman and Hall/CRC Press.

Taylor, J. W., & Buizza, R. (2002). Neural network load forecasting with weather ensemble predictions. *IEEE Transactions on Power Systems, 17*(3), 626–632.

Tan, P. N. et al. (2018) *Introduction to data mining* (2nd ed.). Pearson.

Tuchong. (2020a). *Components of data preprocessing.* Retrieved May 6, 2020, from https://stock.tuchong.com/image?imageId=430505720620777839.

Tuchong. (2020b). *Data cleaning.* Retrieved May 6, 2020, from https://stock.tuchong.com/image?imageId=446216436108231091.

Tuchong. (2020c). *Facial landmarks.* Retrieved May 6, 2020, from https://stock.tuc hong.com/image?imageId=255770877421158487.

Tuchong. (2020d). *Data mining from different perspectives.* Retrieved May 6, 2020, from https://stock.tuchong.com/image?imageId=458561529326600579.

Wohlmuth, B. I. (2001). *Discretization methods and iterative solvers based on domain decomposition* (Lecture Notes in Computational Science and Engineering (17), Kindle edition). Springer.

5

Computer Vision

A lot of the future of search is going to be about pictures instead of keywords. Computer vision technology is going to be a big deal.
Mr. Ben Silbermann (CEO of Pinterest, born 1982)

Abstract How do computers see our world? The emergence of AI in 1950s prompted scientists to build machines imitating human in two aspects: How we think and see? The truth is: Human excel at thinking and learning than the machine (computer). We exercise various thinking and learning techniques to acquire knowledge and intelligence. Human has another unique vision ability over the machine. Any computer scientist will tell us that building a machine (robot) to imitate 100% human vision capability is almost an impossible task. This chapter compares the human visual system with computer vision. By imitating human vision, we introduce 3 computer vision components: (1) Figure-scene segmentation; (2) Object recognition, and (3) 3D & VR modelling. After that, we study various latest computer vision technologies and applications related to daily activities.

How do computers *see* our world?

The emergence of AI in 1950s prompted scientists to build machines imitating human in two aspects: How we *think* and *see*? Why?

R. S. T. Lee, *Artificial Intelligence in Daily Life*, https://doi.org/10.1007/978-981-15-7695-9_5

119

The truth is: Human excel at *thinking* and *learning* than the machine (computer). We exercise various thinking and learning techniques to acquire knowledge and intelligence.

Human has another uniqueness *vision* ability over the machine. Any computer scientist will tell us that building a machine (robot) to imitate 100% human vision capability is almost an impossible task. To illustrate how amazing human vision is, try a simple experiment. Just close your eyes for 10 s and open it again, then count how many seconds you need to recognize all things and objects you see inside your room. You will be amazed by how rapid and effortless for a human to recognize objects from a scene, such computer vision process is so-called *scene analysis* and *object recognition* which is a major research topic in AI and computer science.

In this chapter, we explore *Computer Vision (CV)*—how a machine (computer) *sees* the world?

To understand computer vision, we must start with human vision. Firstly, we will study how human sees the world—our visual system compares with computer vision. Secondly, by imitating human vision, we will introduce three main CV components: (1) *figure-scene segmentation*; (2) *object recognition*, and (3) *3D & VR modelling*. Thirdly, we will study active vision on robot vision application with major challenges and technologies. After that, we will explicate how various latest computer vision technologies can be applied in our daily activities.

5.1   What is Computer Vision?

Computer vision (CV) (Szeliski 2010; Forsyth and Ponce 2010) is a field of computer science enabling computers to *see*, *identify*, and *process* images just like human vision. It is alike imparting human intelligence and instincts to a computer but is an extremely difficult task enabling computers to recognize images of different objects in reality.

Computer vision is linked closely with AI. Its goal is not only to see but also to provide appropriate analysis based on the observation and then perform or act accordingly. A good computer vision system should not only capture an object's 2D image it *sees* from the environment, but it must convert it automatically into a 3D object and be able to track its motion. For example, when we design an auto-driving vision system, the vision system not only needs to *see* all vehicles around us, but more importantly it has to

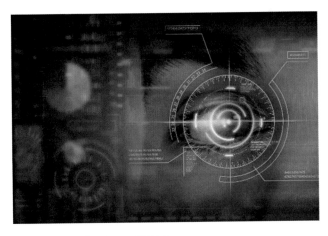

Fig. 5.1 Computer vision (Tuchong 2020a)

know how they currently move and *predict* their future moving speeds and directions in order to navigate our car to avoid any potential collision. It also needs to distinguish all other surrounding objects such as pedestrians, traffic lights, traffic signs, and all possible obstacles to avoid. Remember, these are all not static pictures and objects, but are 3D moving objects in real time.

One might see, a good vision-enable car navigation system to include many capabilities such as: figure-ground segmentation; object recognition; motion checking; motion prediction; anti-obstacle; and real-time navigation.

For instance, such intelligent vision-enable navigation system must provide inputs to the driver or even stop the car automatically if it meets a sudden obstacle on the road (e.g. a dog suddenly runs out to the driving lane), the system must react instantly. It has a similar human vision to identify an object, process data, decide what to do, and complete a complex task in a split-second. CV's aim is to enable computers to perform the same kind of tasks as per human's efficacy (Fig. 5.1).

5.2 How Human See the World?

CV aim is to build an intelligent system to imitate the human vision system. The first basic question is: *How human see the world?*

This question seems to be a biological and physiological problem, but the truth is a lot more. Any high school student undertook human biology knew how we see any external object is based on our visual system: *We can see any external object because light reflects from the external object, e.g. apple enters the lens of our eyes. Alike typical optical system, these lights focus onto the retina of*

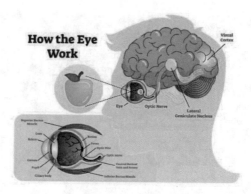

Fig. 5.2 Human visual system (Tuchong 2020b)

our eyes and stimulate our light sensing cells (rods and cones). The optical signals will be sent via the optical nerve into the visual cortex in our brain and then we can see the object. Figure 5.2 illustrates the basic mechanism of how we see an object, e.g. apple via our visual system.

Is this the whole story of how we *see* or *recognize* our world?

The answer is *no.* Actually, this visual system mechanism we all learnt in high school could only tell us how the visual information from an external object captured by our visual system (Remington and Goodwin 2004) and sent to the visual processing center—the visual cortex in our brain. But it did not tell us many important things such as: *How can we (1) separate the object we see from the environment? (2) recognize the object we see? (3) shape and model our visual world?, and (4) navigate in this visual world?*

The first question is what we called *figure-ground segmentation* problem. It seems to be simple and instinctive, in fact, it is a complex and sophisticated process that took millions of years on human and animal evolution to perfect such capability. The second question is not only a visual processing problem but is a rather high-level epistemological problem—the problem and theory of knowledge. The truth is: To *recognize* any object, we must have the knowledge of that object first. In other words, we must have a knowledge base to contain all different objects so that we can recognize that object when you *see* it.

5.3   Real-World Versus Perceived World

The third question is uncanny yet interesting. Have you ever thought about a fundamental phenomenon: *Does the world we see daily is the same world others see?* The answer seems to be *yes,* but it is obviously *no.* A simple argument can answer this question: We all should agree that the world we *see* is based on our visual system. However, the physiological structure of every human cannot be identical, so obviously, the same object we see cannot be completely the same in terms of color, shape, and all other features. In that case, the world we perceive daily maybe similar, but cannot be the same.

It is, in fact, a complex cognitive science or even philosophical problem.

In layman's term, the world we *see* is a 3D model built in our conscious mind so-called *model world* (or *perceived world*). It is a model world we construct subconsciously in our mind based on daily experiences and interactions with the *real-world* via our perceptual system. This information about the real-world comes to us initially through our sensory system: vision (eyes), hearing (ears), smell (nose), taste (mouth), and touch (skin).

According to the Greek philosopher Plato's book *The Republic*, the story of *Allegory of the Cave* (Plato 2017), described our perceived world is like a *projection* of the real-world, similar to 3D adventure games or VR (*Virtual Reality*) games we play in cyberworld (Magnor and Sorkine-Hornung 2020), or a world expressed in the remarkable sci-fi movie *Matrix* in 1990s (Wymann 2002). Once we identified what we *see* is just the 3D model we built inside our mind, how we navigate and interact with this world is nothing more than playing an interactive VR game! Because of this new understanding, some *new age* theory such as *Simulation Theory* proposed that we live in a world of simulation, all perception we have include our sensory system are only electrical signals and information fed by the so-called *construct*, we all interact within this simulated world and assume it is the *reality*. No one knows whether it is true. However, it is a good example to understand how human perceive the world, and how surprising close to our perceived world as compared with the one simulated by intelligent computer systems (Fig. 5.3).

Fig. 5.3 VR game (Tuchong 2020c)

5.4   How Computer See?

Once we learnt how human vision works, the next question is how to imitate human vision into computer vision. Let us begin to look at the basic visual sensing differences of human versus computer.

In the previous section, we studied how human *see* objects, so how about computer vision? Thanks to theoretical physicist Prof. Albert Einstein (1879–1955) for his discovery of *photoelectric effect* (Einstein and Bruskiewich 2014) in 1905, which awarded him the Nobel Prize in 1921, for his discovery of photoelectric effect instead of *General Relativity*. His *photoelectric experiment* as illustrated in Fig. 5.4, showed that light beams shined on a metallic plate trigger electrons emission and received at the other end of the vacuum tube to drive current flow in the electric circuit. This innovative experiment set the cornerstone for light sensor R&D on digital cameras and mobile phones.

The core of every digital camera has a solid-state device (called *light sensor* or *image sensor*) based on the photoelectric effect to capture light entering the lens to form image pixels.

Figure 5.5 illustrates how CV operates via *digital image sensor* (Nakamura 2005). As one might see, the basic visual processing mechanism inside a digital camera is alike the interior of our eye, the major difference is that digital camera uses millions of image sensors in the form of massive 2D-array to convert incoming light into electronic signals and 2D images. Although it seems that one is an electronic device and the other is a purely biological organ. The truth is: these two *devices* convert the incoming light into electrical signals ultimately . The only difference is that in the human visual

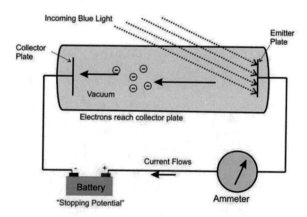

Fig. 5.4 Einstein's famous photoelectric experiment

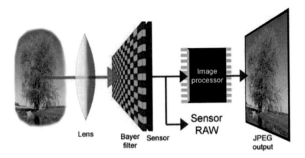

Fig. 5.5 Computer vision via digital image sensor

system, electrical flows are conducted via our visual nerves to the visual cortex inside our brain.

5.5   Main Components of Computer Vision

Like human vision, visual sensing of the environment—or what we called the *scene* in computer vision perspective is only the beginning of the story. To imitate human vision, CV needs to solve the following problems (Bhuyan 2019; Forsyth and Ponce 2010):

1. Segment object from the environment—figure-ground segmentation
2. Recognize object from the scene—object recognition
3. Model and navigate inside the computer world—3D & VR modelling

These three domains of computer vision are also the three main components of the CV. Its R&D is mainly focused on fundamental visual-related problems such as image processing, image transformation, and object recognition in the past 50 years. However, owing to the advance of computer technology, digital image sensing and robotic technology, the scope of CV is extended to a more comprehensive and practical level that is closely related to our daily activities. These examples include:

1. How to use biometric technology such as human face, fingerprint, palmprint or iris for entry control and user authentication?
2. How to recognize terrorists inside the public area such as airport and train station using *gait recognition* (the way and gesture of walking)?
3. How to design and navigate a drone to deliver parcels from suppliers to customers using both GPS and active vision technology?

4. How to design and navigate cars in next-generation of intelligent city using 5G + visual-based auto-driving and navigation technology?

5.6   Figure-Ground Segmentation

5.6.1   Gestalt Theory of Visual Perception

Before we explore how the computer performs figure-ground segmentation, let's investigate how human achieve that goal. Figure-ground segmentation has a long history of study not in computer science, but in *visual perception psychology* (Hamlyn 2017)—a branch of psychology focus on how we percept our world in terms of visual perception to be considered as the original computer vision of psychology.

Hence, *Gestalt psychology* (or *Gestaltism*) is one of the most influential theories in visual psychology emerged in early twentieth century based on the works by psychologists Profs Max Wertheimer (1880–1943), Wolfgang Köhler (1887–1967), and Kurt Koffka (1886–1941). Gestalt Psychology explained how to acquire and maintain meaningful perceptions in our world of perception (Kohler 2014; Wertheimer et al. 2012).

The central principle of Gestalt psychology is that the mind forms a global whole with self-organizing tendencies. This principle maintains that when the human mind (perceptual system) forms a percept or *gestalt*, the whole has a reality of its own, independent of the parts.

The key principles of gestalt systems are:

- Emergence—human mind auto-organize *parts* to the *whole*.
- Reification—constructive or generative aspect of perception (human mind complete *gaps* to form meaning).
- Multi-stability—the tendency of ambiguous perceptual experiences to pop back and forth unstably between two or more alternative interpretations.
- Invariance—the property of perception whereby simple geometrical objects recognized independent of rotation, translation, and scale.

Figure 5.6 shows typical examples found in Gestalt psychology on visual perception providing a distinguished explanation of human perform figure-ground segmentation to recognize the *whole object* from its *parts*. The most famous phenomena of Gestalt psychology are *rubin-vase* and *old-lady/young-girl figure-ground segmentation* shown in Fig. 5.6a, b. Human

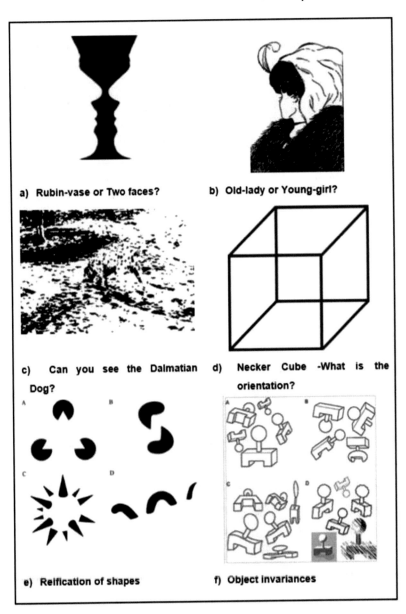

a) Rubin-vase or Two faces?

b) Old-lady or Young-girl?

c) Can you see the Dalmatian Dog?

d) Necker Cube -What is the orientation?

e) Reification of shapes

f) Object invariances

Fig. 5.6 Gestalt psychology on visual perception (Kohler 2014)

can perform segmentation-ground segmentation subconsciously to recognize either a rubin-vase or two human faces but not both simultaneously. Why?

In regard to Gestalt psychology, we have inherited the ability to *extract* an object (from our mind) out of a scene, the object should be recognizable in our mind and the scene is not recognizable (or meaningless in context) as shown in Fig. 5.6c. The importance is that it not only demonstrated the phenomena of how human perform figure-ground segmentation, but also various object recognition problems with invariant properties in Fig. 5.6f.

5.6.2 Traditional Figure-Ground Segmentation Methods

Traditional CV methods on figure-ground segmentation are mainly focused on four areas: (1) color threshold; (2) edge detection; (3) shape detection; and texture detection (Bhuyan 2019; Forsyth and Ponce 2010).

Color Threshold

Color threshold is based on the simple fact that a figure object should be in different color, normally from the background scene. By using color threshold and analysis methods, we can separate the figure object technically from the background scene. Figure 5.7 shows how color threshold method works to extract a scene of the colored fruits by using *MATLAB color threshold toolbox* (Gonzalez 2009).

The Color Thresholder toolbox displays an image in the *Choose a Color Space* tab, with point clouds representing the image in these color spaces:

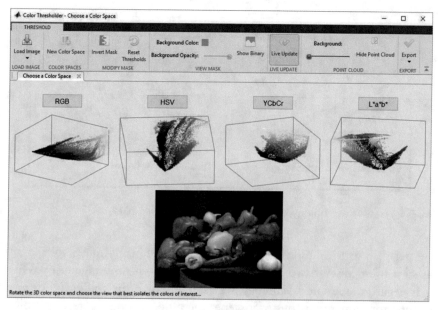

Fig. 5.7 Figure-ground segmentation using Color Threshold method by MATLAB

RGB, HSV, and YCbCr. For color-based segmentation, select the color space that provides the best color separation. Use the mouse to rotate the point cloud representations to see how they isolate individual colors.

Segmentation using *Color Thresholder* can be an iterative process—attempts on several different color spaces to achieve a fitting segmentation. The advantages of color threshold are simple to use and understand. This method can only be applied to simple objects with unique colors but not on complex objects with complex texture.

Edge Detection

Edge detection (Bhuyan 2019; Channar 2020) is one of the most frequently used figure-ground segmentation techniques in computer vision. It derived from the basic concept that any figure object has its own shape and boundary (called *edge*) from the surrounding scene. We can segment figure object(s) from the scene and environment by using simple edge-detection functions. Techniques frequently used include: Canny, Sobel, Prewitt, and Roberts methods. The advantages of edge-detection techniques are that they distribute many graphical and scientific project development applications as standard image processing libraries and functions such as OpenCV (Kaehler and Bradski 2016) and MATLAB (Gonzalez 2009). Figure 5.8 illustrates figure-ground segmentation using the canny edge-detection method on the famous *Lena* picture generated by MATLAB image processing toolbox. The major disadvantage of edge-detection method for figure-ground segmentation is that such a method is only useful in two conditions: (1) The background scene is not too complex. If it is and full of other pattens, it is easy to segment the wrong object from the background; (2) Only one or two figure object(s)

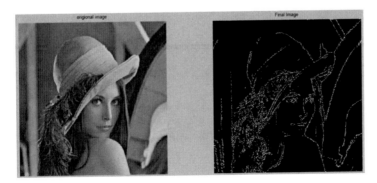

Fig. 5.8 Figure-ground segmentation using Canny edge-detection method on Lena generated by MATLAB image processing toolbox

appeared in the scene. In other words, if there are many figure objects in the background scene, it is difficult to segment all figure objects effectively.

Shape Detection

Shape detection (Bhuyan 2019; Peters 2017) method derived from the concept that figure-object should either: (1) has a well-defined shape or (2) has a completely different texture from the surrounding environment. By distinguishing these *standard shapes* or *textures* from the environment, we can segment the figure object technically from the scene. Frequently used shape detection techniques include *contour, elastic graph,* and *snake tracking methods*. Simply speaking, contouring method borrows the concept of contour drawing for landscapes and mountains by fitting some regular shapes, e.g. triangles, square, rectangle, or polygons into the scene to extract figure object(s). Snake and elastic graph tracking methods extend the idea by using a snake or elastic graphs (is, in fact, a set of mathematical formulae) to trace the contours and extract figure object(s). Again, these methods are frequently used and now become a standard library for many image processing and data mining tools such as *MATLAB, OpenCV* even *Python* tools (Solem 2012). Figure 5.9 demonstrates how a snake model works to extract figure object from the clustered scene with a variety of objects include: a human at the foreground, a monitor and PC at the back. Also, there is a bird at the right-bottom corner. Can you spot it ?

Fig. 5.9 Using snake method to extract figure object from the clustered scene

Fig. 5.10 Texture-based object segmentation using *Gabor Filter*

Texture Detection

Texture detection (Bhuyan 2019) method derived from the concept that the figure object should have a completely different texture as compared with its surrounding environment. By filtering-out the background by texture analysis, we can technically extract figure object(s) from the scene accordingly. Frequently used texture analysis schemes are texture filter using various filtering techniques such as *linear filtering, anisotropic filtering,* and *mipmap filtering techniques* that integrated with several image processing tools such as *MATLAB* and *OpenGL*. In fact, the most popular texture filtering method used commercially is *Gabor Filter* to signal processing and noise filtering applications.

Figure 5.10 illustrates an example of the texture-based object, e.g. a dog segmentation using *Gabor Filter* performed by MATLAB image processing toolbox (Gonzalez 2009). As said, texture-detection method is good when either the object or background scene has a completely different texture. If both background and figure object do not have well-defined texture, they will deteriorate the performance of the texture filter significantly.

5.6.3 Neural Oscillators in Our Brain

In the previous chapter, we studied the biological neural network of our brain. Current neurophysiology advises us that all neurons of our brain exist in the state of *oscillations, chaotic oscillation* to be exact. The latest neuroscience studies even reveal that our *thinking, learning,* and possibly vision processing is achieved in the form of neural oscillations (Lee 2006a).

Can we *see* them?

The answer is: We observe and realize their existence long ago but not aware of how important they are. The most common example is called *brain waves*—the *EEG (electroencephalogram)* patterns of our brain at different stages of our daily activities which are collective neural oscillation patterns of neurons inside our brain. There is a total of five brain waves types (Someren et al. 2011) appear in different stages of our awareness as shown in Fig. 5.11:

- Beta wave (16–30 Hz)—awake, conscious state.
- Gamma wave (31–100 Hz)—insight and peak focus conscious state.
- Alpha wave (8–15 Hz)—relaxed, calm, lucid dream (but not thinking).
- Theta wave (4–7 Hz)—REM dream state, meditation, deep relaxation.
- Delta wave (0.1–3 Hz)—deep, dreamless sleep.

Additionally, neuroscientists today even believe that such neural oscillation exists in the form of *chaotic oscillation*. They not only exist in brain but also found in all our five senses of perception (Arbib et al. 1991):

- Visual nerve—eye (vision)
- Olfactory nerve—nose (smell)
- Auditory nerve—ear (sound)
- Gustatory nerves—tongue (taste)
- Somatosensory nerves—skin (touch)

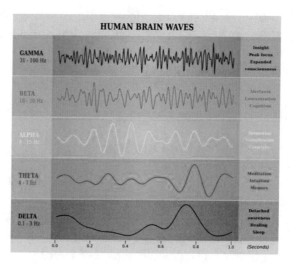

Fig. 5.11 Human brain wave patterns

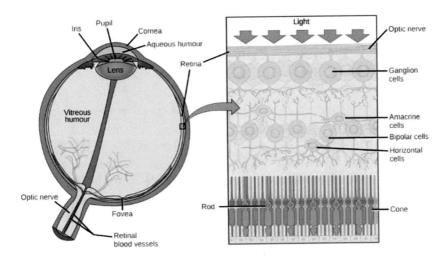

Fig. 5.12 Neural structure of visual processing

Figure 5.12 illustrates the neural structure of visual processing.

5.6.4 Figure-Ground Segmentation Using Neuro-oscillators

Current neurophysiological evidence revealed that the neural interaction of our visual process is a typical photo-electrical effect between light wave (photons) and neurons inside our optic nerve in the form of *chaotic neural oscillations*. The frequency of these oscillations is in the range of 40 Hz (i.e. Gamma wave) and differed from the periodic activation induced by grating, suggesting that the oscillations and their synchronization were due to internal neuronal interactions, providing extensive evidence of the functional role of gamma oscillations in visual perception and feature binding (Basar 2007).

By using this new theory, neuroscientists believe that how we can segment a figure object from a complex scene in a just split-second (maybe less than 1 s) is not because we use traditional *edge-detection* or *texture-detection* schemes, but instead are rapid and simultaneous neural oscillations once visual neurons contact with visual stimulus and took few time-cycles only for the figure object to *popup* onto your conscious mind (Lee 2003, 2006a).

The problem is, how can we construct such a powerful neural oscillator?

After over 5 years of research and numerous tests, the author created an ideal chaotic neural oscillator, now known as *Lee-oscillator* in 2004, and published in two major neural network journals *IEEE Transactions on Neural Networks* (Lee 2004), for the introduction of Lee-oscillator with its transient chaotic behavior; and *Journal of Neural Networks* (Lee 2006a), for the exploration of Lee-oscillator on the stimulation of human visual perception with *progressive memory recalling mechanisms.*

Figure 5.13 depicts the neural model of Lee-oscillator.

Basically, Lee-oscillator consists of 4 neural elements: E, I, Ω, and L which corresponds to *excitatory, inhibitory, input, and output neurons,* where e1, e2, i1, and i2 are the weights and S(t) is the external input stimulus.

Being an analogue of the classical *Hopfield network* as an auto-associator, a transient chaotic auto-associative network based on Lee oscillators as its constituting neuron elements can be used to provide an innovative progressive memory association and recalling scheme (Lee 2004, 2006a) to expand how human can recall memory and to explain the important phenomena—*Gestalt visual psychology* on figure-ground segmentation.

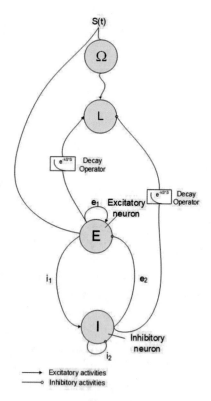

Fig. 5.13 Lee-oscillator

5.6.5 Lee-Associator and Gestalt Visual Perception

The direct adoption of Lee-oscillator is based on a simple 2D single-layered neural population analogous to a classical Hopfield network studied in the previous chapter. A collection of Lee oscillators as basic neural elements is constructed to form the Lee-associator (Lee 2006b). Figure 5.14 illustrates a schematic diagram of the Lee-associator.

The interactions among constituent neurons of Lee oscillators in this network can act as an auto-associator in the presence of query patterns that are treated as external input stimuli, in analogy to the classical Hopfield network. However, in contrast to these classical models, the proposed Lee-associator provide a change in neural dynamics (from a chaotic to a stable state transition) when pattern association occurs with remarkable progressive memory construction (recalling) and Gestalt visual psychology on figure-ground segmentation. Figure 5.15 shows how Lee-associator uses chaotic neural oscillation method to perform fast figure-ground segmentation on the famous rubin-vase test of Gestalt visual psychology (Lee 2006b).

The figure on the left showed a standard visual stimulus with the complete rubin-vase object. As shown in the oscillation chart (left), within 50 time-cycle, Lee-associator only took 6 time-steps to complete the vase object extraction. In terms of gamma wave with the visual perception of 40 Hz, 6 time-steps of figure-ground segmentation rate corresponded to 6/40 =

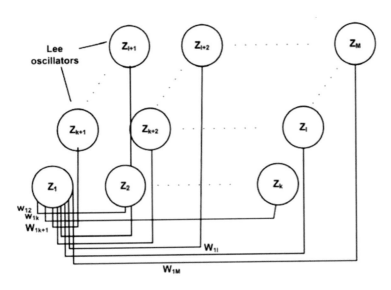

Fig. 5.14 Lee-associator

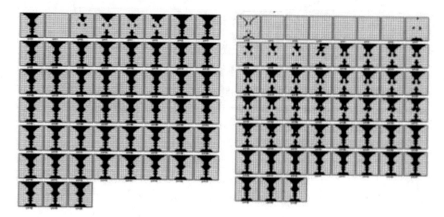

Fig. 5.15 Rubin-vase experiment using Lee-associator (left: original rubin-vase; right: rubin-vase shape)

0.15 s, which are evenly matched to human figure-ground segmentation rate on *rubin-vase experiments*.

Another important evidence revealed from the *rubin-vase experiment* (Lee, 2006a) was that the Lee-associator could truly extract either *rubin-vase* or *two human face profiles* during the experiment, mainly due to the fact that the Lee-associator performed chaotic oscillations in each test. In fact, the results could oscillate between 2 possible outcomes, alike human performance in visual perception. The figure on the right revealed another important property of Lee-associator in figure-ground segmentation with the provision of the partial shape of the figure object. As shown in experiment results, it took around 36 time-steps for Lee-associator to *figure out* the rubin-vase with the provision of a partial shape. From the human perspective, it corresponded to around 0.9 s for figure object extraction.

In fact, chaotic neural oscillators can not only be applied to figure-ground segmentation, but also to other object recognition and visual tracking problems.

5.7   Objection Recognition

5.7.1   How Human Recognize Objects?

Let's take a look at Fig. 5.16. What object(s) do you see? A dog, a plant, some leaves, or nothing?

Fig. 5.16 What object you are seeing? (Tuchong 2020d)

If you see a dog. What kind of dog it is? Is it your dog? If not, why are you so sure? As we can see, a simple picture like this can tell us a lot about how powerful human handle object vision and recognition problems (Cyganek 2013; Lee 2003). Once we see such picture, only within a split-second, we can perform consciously or even subconsciously all these vision processes in seconds (or less) which include (Bhuyan 2019):

1. Figure-ground segmentation of figure object—*extract* the *dog* image from background scene.
2. Object recognition—*mentally match* this *dog* object from our memory and recognize it as a dog.
3. Knowledge acquisition—*retrieve* related information and knowledge about this *dog* from our memory.

We already learnt figure-ground segmentation in the previous sections. In fact, some individuals working on image processing nevertheless believe that processes 1 and 2 are identical. Their argument is that: In terms of image processing, is it technically possible to recognize an object from a picture with the complex background?

The answer is *yes* and *no*. *Yes,* in the sense that in terms of image processing, we can do so. But if we wish to implement CV, that is, to simulate how human visual processing and recognition process, they will be separate steps and processes. For example, when we enter a classroom with over 40 students, what we *see* in a split-second is not a *2D picture of the classroom with students,* but possibly a *classroom of many students' faces.* Latest visual and cognitive psychology revealed that humans are particularly skillful in segment human faces for recognition. In terms of machine learning and neural networks, our brains are somehow *hardwired* to extract human face from a complex scene and recognize it at an astonishing speed from out of thousands of faces we came across! Once we recognize the object, human will by default subconsciously relate it to a lot of attributes which include: names, sensations, events, images even videos that are all belonged to an important concept and theory so-called *ontological knowledge acquisition*. We will study this in Chap. 7. First, let us take a look at how the machine recognizes objects (Lim et al. 2011).

5.7.2 Classical Object Recognition Model

Over the years, there are numerous CV techniques on *object recognition*. However, all object recognition systems aim to solve a fundamental problem: Out of a collection of memory objects (may be thousands or millions), how to design a computer program to examine a match with a tested (or unseen) object?

This is a two-step process: (1) information extraction module; (2) object matching module as shown in Fig. 5.17.

Information extraction module aims to extract vital information from figure object for recognition. *Object matching module* derives information extracted from the figure object and compares them with all memory objects stored in the memory database in order to find the best match (within a certain threshold level), or *no match* if the best match cannot be found from the memory database. In that case, the figure object can be stored as a new memory object in the memory database as a new *memory*.

In terms of object information extraction, although there are other methods and techniques developed in the past decades, they can be categorized into three main approaches (Cyganek 2013; Lee 2003):

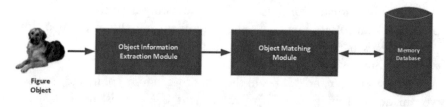

Fig. 5.17 Classical object recognition model

- Model extraction technique.
- Feature extraction technique.
- Combined extraction technique.

5.7.3 Model Extraction and Object Matching Technique

Model-based extraction technique derives from the concept that every object has its own shape and construction. By modeling their shape and construction, we use their information as vectors for pattern matching with memory objects stored in object databank for object recognition. Frequently used methods range from simple *polygon modeling, contour, and snake models* studied in the previous sections on figure-ground segmentation.

However, these methods are used to extract the object model with simple structure. For figure object such as human faces with complex shapes and distinctive features, a high-level model-based system such as *elastic graph matching* (EGM) method is frequently used in human facial recognition system nowadays. In addition to usual tricks to extract the facial shape (contour), there are two keys features for facial mask using EGM technique (Lee 2003):

- Distinctive facial points known as *facial landmarks* are used to create a 2D/3D graph of a human face. These facial landmarks include all vital facial points such as forehand, eyebrows, eyes, nose, cheekbones, mouth, lips, chin, etc.
- 2D/3D is constructed in the form of an *elastic graph* in order to provide invariance properties for different facial expressions so that it can recognize the same person even though with different expressions such as smiling and laughing, or other poses.

Figure 5.18 shows a typical model-based extraction technique using elastic graph matching method (EGM) of a human face. In terms of object matching, it depends on the model extraction method being used. For facial mask modeling method, the elastic graph matching algorithm can be applied directly for object matching and recognition. The major advantage of model-based extraction technique such as EGM method is that if all landmarks are well-defined, the model extraction rate and object recognition rate is quite high with up to 85–90% accuracy. However, if the figure object itself contains complex feature patterns, model-based extraction and object matching would not be able to apply effectively.

Fig. 5.18 Elastic Graph Matching on human face (Tuchong 2020e)

5.7.4 Feature Extraction and Object Matching Technique

The feature-based extraction technique is different from the model-based extraction technique. It is believed that pattern features of figure object are the key information for identification. By the extraction of pattern features and compared with the objects stored in the object database, we could perform object recognition effectively. What are pattern features?

In fact, pattern features can be a range from low-level features such as object color or color histogram, texture patterns studied in the previous section, to high-level features extracted by algorithms (programs) such as *Gabor features*, or *Gabor wavelets* used in human face feature extraction scheme (Lee 2003; Nixon and Aguado 2019). Figure 5.19 shows how feature-based extraction is used on human face feature extraction.

Fig. 5.19 Feature-based extraction technique using Gabor wavelets on human faces (Tuchong 2020f)

In this example, we used *Gabor wavelets* to extract high-level facial features from 39 facial landmarks of a human face.

What is *wavelet*? Why it is so powerful?

The concept is simple. *Wavelet* is a kind of localize wavefunction with limited *length* to model and represent a feature point as *multidimensional feature vectors*. If we extract wavelet features of 39 facial landmarks of a human face, it becomes a digital ID with 39 attribute vectors like our fingerprint. Gabor wavelets are used to extract feature vectors with over 100–150 feature points in a typical fingerprint in many fingerprint recognition systems nowadays.

Owing to its reliability and high accuracy, feature-based facial landmark extraction and human face identification system are used by government agencies to identify potential suspects and terrorists over decades, and are frequently used by many governments and commercial faculties for biometric-based entry and security control.

However, the accuracy depends highly on the correct identification of feature landmarks. If feature landmarks cannot be clearly defined, or even not exist, it would not be able to apply.

5.7.5 Combine Extraction and Object Matching Technique

In short, the combined extraction technique is an integration of *model* and *feature extraction techniques*. It is frequently used in real-time human face recognition system in many governments and public facilities such as airports and train stations for entry authentication and access control.

The combination of *model extraction technique* is a frequently used method using elastic graph matching (EGM) method and *feature extraction technique* with *Gabor wavelet feature extraction method* on human face recognition (Lee 2003). As mentioned, the EGM method is good with object matching by using its *inherited* elastic graph property. Therefore, it can be used as the first step to *mask* the human face and identify the 39 facial landmarks. The intrinsic problem of feature-based extraction technique would be solved automatically for correct facial landmark positions identification. Once these facial landmarks are clearly located, then *Gabor wavelets feature extraction techniques* can be applied to extract all wavelet features of the 39 facial landmarks. Figure 5.20 shows the combined extraction technique using *EGM + Gabor wavelets* on human faces.

One might ask, why don't we use the combined methods at the very beginning?

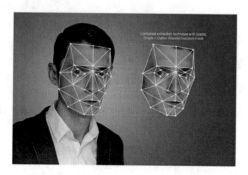

Fig. 5.20 Combined extraction technique on human faces (Tuchong 2020g)

The truth is, to build a real-time facial recognition system, says security and entry control in an airport, the average facial recognition processing time is a critical issue. In terms of object recognition speed, feature extraction is not a major issue. However, the elastic graph matching method is an intensive computational task. But thanks to the advance computer technology, big data development, and cloud computing systems; the overall real-time facial mask and feature extraction speed improved substantially in the past 10 years. Together with powerful storage and computational capacity of the facial database stored in a cloud system, real-time facial recognition system support over thousands or up to millions of users become biometric-based entry control and security systems.

5.7.6 Objection Recognition Using Neural Oscillators

We had studied how neural oscillators (Lee oscillators) resolve Gestalt visual perception for the extraction of figure object effortlessly (Lee 2004, 2006a) in the previous section on figure-ground segmentation. Can neural oscillators be applied to object recognition?

The answer is definitely *yes*. Besides the resolution of Gestalt visual perception on figure-ground segmentation of rubin-vase experiment presented in the original paper of Lee-associator (Lee 2006b), an important experiment to demonstrate how Lee-associator can be applied to complex object such as human face recognition. In the test, the author adopted the *Yale University facial database (set A)* which contained facial images of over 3000 human faces, each individual contains 10–15 poses from different views and facial expressions including frontal, side views, gimmick, and occluded faces as in Fig. 5.21a. Figure 5.21b depicts sample facial patterns from the database and *progressive memory recalling (PMR)* of the Lee-associator on recalling facial

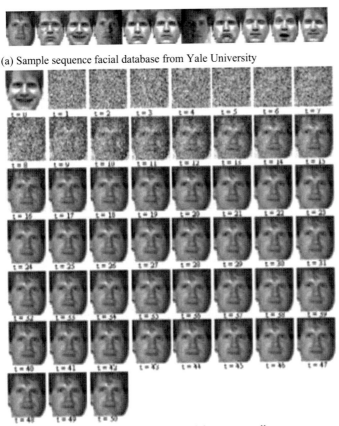

(a) Sample sequence facial database from Yale University

(b) A sample sequence of a progressive facial pattern recall

Fig. 5.21 Progressive memory recall of human faces by Lee-Associator (Lee 2006b)

patterns (Lee 2006b). As one can see, it is different from traditional mode extraction or feature extraction and object recognition scheme. Lee-associator demonstrates a different way of object recognition method by some sort of memory recalling scheme, from complete noise pattern to *reconstruct* the correct human face stored in memory storage that resemble how human *recall* a human face from own memory instead of algorithmic matching and distance evaluation in classical computer vision methods. In terms of object recognition rate, Lee-associator took around 20 time-steps to recall a correct human face. By adopting human's thinking and perception conscious stages with Gamma brain wave of around 40 Hz, 20 time-steps corresponded to around 0.5 s for success matching, which was quite sensible in real-life scenario to recall a stranger's face from our memory. The research of neural oscillators on CV was, in fact, only the beginning. There had been other research being conducted at different areas ranging from object recognition and recalling, to see how we interrelate different ideas and concepts inside our minds—*ontology* and *knowledge base* to be studied in Chap. 7.

5.8   3D & VR Modelling

5.8.1   VR and Shared Consciousness

What is the world we see? Is it a true reality, or a one we imagine and create by our minds?

The first time the author came across this problem was heard from the story about Plato's *Allegory of the Cave* in high school (Plato 2017). In this allegory, humans are only *prisoners* blinded by own mind as in Fig. 5.22. The world we see and live is a *projection* of the world. It is so *real* and *touchable* that we believe it to be the true world and called it the *reality*.

The author was fairly puzzled by this idea at the time until the research on AI and robotic vision in 1997. From hia Ph.D. study on AI-based object recognition to implement own designed CV system on robotic vision, it came up with an interesting question . To allow robots *see*, *navigate*, and

Fig. 5.22 Plato's *Allegory of the Cave* (Tuchong 2020h)

Fig. 5.23 VR and shared consciousness (Tuchong 2020i)

interact with the surrounding world, only figure-ground segmentation and object recognition were not enough. The first thing we need is to construct a *reality* (or *virtual reality, VR* we called nowadays) for the robot to *live in,* to *navigate,* and to *interact,* like what we play with the 3D/VR adventure games. From robot's perspective, because VR is part of the construction of the complete robotic system, the robot has no way to differentiate whether it is the true reality or not, as Plato mentioned in his allegory of prisoners inside the cave.

Take a step further, if the world we see is only a VR, how about our *worlds* we all interact? Are they the same reality, or each of us has our own reality?

This problem seems to be somewhat philosophic, but if we have experiences in playing multiple player VR-based adventure game, this would not be a problem. We are represented by our *avatars* inside the VR world (Guazzaroni and Pillai 2019). This VR world is a *construction* or *shared consciousness* we live and interact with each other as in Fig. 5.23, like what we saw in remarkable sci-fi movies such as *Matrix* and *Avatar.* The question is, how can we construct such a 3D & VR world?

5.8.2 3D Modelling Technology in Computer Vision

To construct a VR world, the first step is to tackle 3D modeling problem – how to build 3D objects models appeared in the VR world? (Guazzaroni and Pillai 2019). This is, in fact, one of the passionate problems in CV and AI technology, not only for academic and research purpose; but also critical for commercial and industrial use ranging from 3D movie production,

3D/VR games design to medical applications such as dentist and surgical robots which is an important component for intelligent health systems to be studied in the next part of this book.

Over the past 30 years, there had been many methods and technologies proposed to build 3D models which can be categorized into three main types (Wöhler 2012):

- *Polygonal modeling*—points in 3D space, called *vertices*, are connected by line segments to form a *polygon mesh*. However, polygons are planar and can only approximate curved surfaces by using numerous polygons.
- *Curve modeling*—a 3D modelling method that relies on curves to generate surface geometry and influenced by weighted control points. Curve modelling can either be parametric (i.e. based on geometric and functional parameters) or freeform (see freeform surface modelling) and rely on nonuniform rational *B-splines* to describe surface forms. This method is also popular in 3D printing.
- *Digital sculpting*—also known as *sculpt modeling* or *3D sculpting*, is a new method to use software that offers tools to push, pull, smooth, grab, pinch or otherwise manipulate a digital object as if it were made from a real-life substance such as clay.

How about if we want to digitize and model a *real* 3D object (a human actor) and try to track his/her motion? Can we do that?

The answer is definitely *yes* and frequently used in film and VR game industry in the past 20 years. The concept is simple as shown in Fig. 5.24. To construct a 3D human object model and perform motion tracking, what we need is to use over 50 video cameras and surround the human object. By doing so, we use 50 2D video pictures from different perspectives (for each frame of the video) to recreate a 3D model of the human object. Also, trackers (in white) are placed onto the joints of the human skeleton to track human object precise motions as shown. The truth is: Almost all 3D/VR movies produced in Hollywood in the past 20 years used the research technology to produce true 3D motion and Matrix was the first movie applied in commercial use (Guazzaroni and Pillai 2019).

5.8.3 3D from VR to AR

Once we have the knowledge and techniques on how to model a 3D object, our next challenge is to construct *virtual reality (VR)*. Virtual reality means

Fig. 5.24 3D modeling and tracking with multiple cameras

creating immersive, computer-generated environments that are so convincing for users to react the same way as they would in real-life. Figure 5.25a shows VR technology in a simulation city. The idea is to block out sensory input from outside and use visual and auditory cues to make the virtual world seem more real (Guazzaroni and Pillai 2019).

To achieve these, a true VR system should fulfill the following requirements:

1. *Convincible*: A true VR world so convincible that we (the person participants in real-world) can hardly differentiate.

(a) VR in a simulation city
(Tuchong 2020j)

(b) AR inside a library
(Tuchong 2020k)

Fig. 5.25 Virtual Reality (VR) vs Augmented Reality (AR)

2. *Interactive*: If we move around inside the VR world, all surrounding objects and environment should move with us seamlessly, so that we can interact with the environment instantaneously without any time delay.

3. *Real-time 3D-simulation*: It is one of the most challenging parts in VR technology. Thanks to todays' computing technology together with the latest 3D modeling technology studied previously in the previous section, real-time 3D-simulation is not a dream anymore.

4. *Explorable*: A VR world needs to be huge in scope and detailed enough for us to explore. It is like playing an adventure game, a true VR world would consist of thousands or even millions of scenes for users to explore and interact, which have substantial memory and CPU capacity.

5. *Immersive*: To be convincible and interactive, VR needs to engage both our bodies and minds. A true VR system must replace our senses of vision, sound and even touch, taste, and smell by certain VR devices such as 3D VR glasses, head-mounted displays (HMDs), data-gloves, motion sensors, etc.

Augmented reality (AR) is an interactive experience of a real-world environment where objects reside in the real-world are enhanced by computer-generated perceptual information, sometimes across multiple sensory modalities including visual, auditory, haptic, somatosensory, and olfactory (Guazzaroni and Pillai 2019). Figure 5.25b shows how AR works in a library. A true AR library should look-and-feel exactly like the real library with value-added functions and services such as interactive search as shown.

5.9   Applications of Computer Vision in Daily Activities

As we can see, the enhancement of computer vision covers figure-ground segmentation and object detection; model and feature extraction; object recognition; object tracking; 3D and VR modeling, and robotic vision. It is different from the past that core technology was used on automated image analysis in many fields at most, while machine vision usually refers to a process of combining automated image analysis with other methods and technologies, to provide automated machine and robotic vision systems for industrial applications. In fact, computer vision overlaps with machine vision (MV) in many application areas significantly nowadays (Bhuyan 2019; Guazzaroni and Pillai 2019).

With the advance of AI technology, both robotic vision and VR/AR become the major areas of applications in CV and AI technology (López et al. 2017; Bhuyan 2019; Guazzaroni and Pillai 2019).

Real-world applications of computer vision include:

- Face detection in cameras.
- Pedestrians, cars, road detection in smart (self-driving) vehicles.
- Vehicle license plate scanners at security checkpoints.
- Terrain detection in drones and airplanes.
- Object 3D scanning to digitize object's physical appearance.
- Augmented reality (AR), mixed reality (MR) or hybrid reality (HR).
- Vision-based inspection system for industry, i.e. oil pipeline defects.
- Robotic vision on space exploration at foreign terrain, i.e. Mars exploration (Fig. 5.26).

CV is a key AI component in many real-world applications of our daily activities. It is functional to use at areas like intelligent campus and city such as CV + GPS-based package delivery by drones as in Fig. 5.27, which will be studied in the next part of this book.

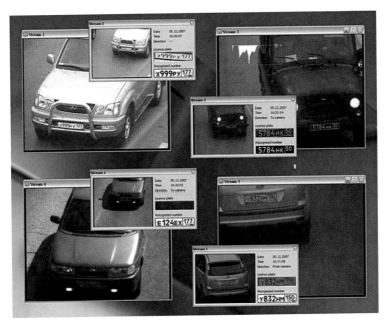

Fig. 5.26 Vehicle license plate scanners at security checkpoints

Fig. 5.27 CV + GPS technology for automatic delivery of package using drones (Tuchong 2020l)

5.10   Case Study: Gait Recognition

Gait recognition (GR) (Lam et al 2007; Basar 2007; Balážia 2013) is a lesser known but a powerful biometric recognition method in addition to frequently used biometrics such as face, iris, fingerprint, and palmprint recognition. In short, GR derives from our *walking manners* for identification as shown in Fig. 5.28.

The theory behind this recognition system is that every individual has a unique gait. It has also been a common experience that a familiar person can be recognized by his/her own *gait* from a distance. The increasing influence of biometrics in today's personal recognition needs has led researchers to leverage gait recognition capabilities. It is one of the few recognition methods that can identify people from a distance and able to improve accuracy when used with other security and surveillance techniques.

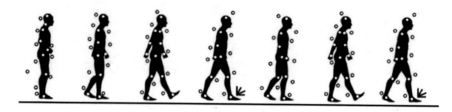

Fig. 5.28   Gait recognition—our walking manners

Carry out basic research on gait recognition and answer the following questions:

1. Why gait recognition is possible as a kind of biometric property in terms of behavior psychology?
2. Today, face and fingerprint recognition has become an industrial standard and widely used in many commercial and government facilities for security control and user authentication. Why do we still need gait recognition?
3. What are the possible application areas and scenarios of gait recognition?

Suppose you are assigned to design a gait recognition security inside the airport for security control:

1. Based on the gait recognition system architecture given in Fig. 5.29, discuss and explain how each module works.
2. Based on the various component of computer vision learnt in previous sections, propose and explain various CV methods that can be applied to each functional module shown in Fig. 5.29.
3. Give two real-world examples, explain how gait recognition system can be used in an intelligent city.

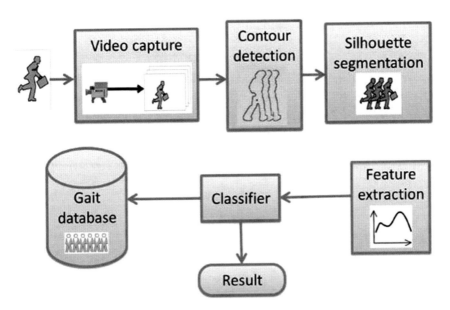

Fig. 5.29 System architecture of a gait recognition system

5.11 Conclusion

We discuss the concept of computer vision (CV) and explain the three major components which include figure-ground segmentation, object recognition, and 3D & VR modeling. We also introduce various traditional and latest AI technology related to CV.

The aim of this chapter is to provide some basic understanding and knowledge of CV. More importantly, how such technology be applied to AI applications that are closely related to our daily activities. With the advance of AI and robotic technology, the scope and definition of CV becomes broader and overlaps with machine vision in various aspects such as robotic vision and drone technology.

CV also evolved and become a cross-discipline subject not only focusing computer technology and image processing, but also in other challenging areas that include:

- The exploration of human vs computer vision in terms of visual psychology.
- Integration with deep learning schemes such as *convolution neural networks* (CNN) on complex object recognition and identification.
- Integration with AI-based data mining and pattern discovery technology such as tumor identification and diagnosis.
- Integration with latest 5G & GPS technology on a location-based system such as package delivery and auto-driving.

CV vision becomes an essential component in numerous AI applications to cover a broad range of functional tasks, ranging from traffic surveillance, security control, user authentication to high-end AI application such as auto-driving, tumor detection, AR and robotic surgery and dentist operations. Figure 5.30 illustrates a scenario of how CV is applied to auto-driving at an intelligent city. These are critical AI applications which will be explored in detail in the next part of this book.

Fig. 5.30 Computer vision in intelligent city on auto-driving (Tuchong 2020m)

References

Arbib, M. A., et al. (1991). *Visual structures and integrated functions* (Research Notes in Neural Computing) (Research Notes in Neural Computing (3)). Springer.

Balážia, M. (2013). *Human gait recognition: Based on body component trajectories.* LAP LAMBERT Academic Publishing.

Basar, E. (2007).*Memory and brain dynamics: Oscillations integrating attention, perception, learning, and memory: Oscillations integrating function and memory* (Conceptual Advances in Brain Research Book 7). T & F Books UK.

Bhuyan, M. K. (2019). *Computer vision and image processing: Fundamentals and applications.* CRC Press.

Channar, N. A. (2020). *A comparative study of edge detection techniques in digital images.* Grin Verlag.

Cyganek, B. (2013). *Object detection and recognition in digital images: Theory and practice.* Wiley.

Einstein, A., & Bruskiewich, P. (2014) *Albert Einstein, and the photoelectric effect* (Elements of Quantum Physics Book 2, Kindle edition). Pythagoras Publishing.

Forsyth, D. A., & Ponce, J. P. (2010). *Computer vision: A modern approach* (2nd ed.). Pearson.

Gonzalez, R. C., et al. (2009). *Digital image processing using Matlab.* GP.

Guazzaroni, G., & Pillai, A. S. (2019). *Virtual and augmented reality in education, art, and museums* (Advances in Computational Intelligence and Robotics). IGI Global.

Hamlyn, D. W. (2017). *The psychology of perception: A philosophical examination of gestalt theory and derivative theories of perception* (Psychology Library Editions: Perception Book 13). Routledge.

Kaehler, A., & Bradski, G. (2016). *Learning OpenCV 3: Computer vision in C++ with the OpenCV library.* O'Reilly Media.

Kohler, W. (2014). *Principles of Gestalt psychology.* Mimesis International.

Lam, T. H. W., Lee, R. S. T., & Zhang, D. (2007). Human gait recognition by the fusion of motion and static spatio-temporal templates. *Pattern Recognition, 40*(9), 2563–2573.

Lee, R. S. T. (2006a). *Fuzzy-neuro approach to agent applications: From the AI perspective to modern ontology.* New York; Berlin: Springer.

Lee, R. S. T. (2006). Lee-associator—A chaotic auto-associative network for progressive memory recalling. *Neural Networks, 19*(5), 644–666.

Lee, R. S. T. (2004). A transient-chaotic autoassociative network (TCAN) based on Lee oscillators. *IEEE Transactions on Neural Networks, 15*(5), 1228–1243.

Lee, R. S. T. (2003). *Invariant object recognition based on elastic graph matching: Theory and applications.* IOS Press.

Lim, E. H. Y., Liu, J. N. K., & Lee, R. S. T. (2011). *Knowledge seeker: Ontology modelling for information search and management: A compendium.* Berlin: Springer.

López, A. M., et al. (2017). *Computer vision in vehicle technology: Land, sea, and air.* Wiley.

Magnor, M., & Sorkine-Hornung, A. (2020). *Real VR—Immersive digital reality: how to import the real world into head-mounted immersive displays* (Lecture Notes in Computer Science). Springer.

Nakamura, J. (2005). *Image sensors and signal processing for digital still cameras* (Optical Science and Engineering). CRC Press.

Nixon, M., & Aguado, A. (2019). *Feature extraction and image processing for computer vision* (4th ed.). Academic Press.

Peters, J. F. (2017). *Foundations of computer vision: Computational geometry, visual image structures and object shape detection* (Intelligent Systems Reference Library). Springer.

Plato. (2017). *Plato collection: The allegory of the cave and dialogues* (B. Jowett, Trans.).

Remington, L. A., & Goodwin, D. (2004). *Clinical anatomy of the visual system* (2nd ed.). Butterworth-Heinemann.

Solem, J. E. (2012). *Programming computer vision with Python: Tools and algorithms for analyzing images.* O'Reilly Media.

Someren, et al. (Eds.). (2011). *Slow brain oscillations of sleep, resting state and vigilance* (Vol. 193) (Progress in Brain Research (Volume 193)). Elsevier.

Szeliski, R. (2010). *Computer vision: Algorithms and applications* (Texts in Computer Science). Springer.

Tuchong. (2020a). *Computer vision.* Retrieved June 8, 2020, from https://stock.tuchong.com/image?imageId=262247069628235853.

Tuchong. (2020b). *Human visual system.* Retrieved June 8, 2020, from https://stock.tuchong.com/image?imageId=427524953186894092.

Tuchong. (2020c). *VR game.* Retrieved June 8, 2020, from https://stock.tuchong.com/image?imageId=415758254631157960.

Tuchong. (2020d). *What object you are seeing?* Retrieved June 8, 2020, from https://stock.tuchong.com/image?imageId=452676350389256332.

Tuchong. (2020e). *Elastic Graph Matching on human face*. Retrieved June 8, 2020, from https://stock.tuchong.com/image?imageId=457460651882447753.

Tuchong. (2020f). *Feature-based extraction technique using Gabor wavelets on human faces*. Retrieved June 8, 2020, from https://stock.tuchong.com/image?imageId=260355858088853636.

Tuchong. (2020g). *Combined extraction technique on human faces*. Retrieved June 8, 2020, from https://stock.tuchong.com/image?imageId=257879405715783752.

Tuchong. (2020h). *Plato's Allegory of the Cave*. Retrieved June 8, 2020, from https://stock.tuchong.com/image?imageId=903401144760270890.

Tuchong. (2020i). *VR and shared consciousness*. Retrieved June 8, 2020, from https://stock.tuchong.com/image?imageId=428610454631350343.

Tuchong. (2020j). *VR in a simulation city*. Retrieved June 8, 2020, from https://stock.tuchong.com/image?imageId=459532784640983095.

Tuchong. (2020k). *AR inside a library*. Retrieved June 8, 2020, from https://stock.tuchong.com/image?imageId=262948747025318013.

Tuchong. (2020l). *CV + GPS technology for automatic delivery of package using drones*. Retrieved June 8, 2020, from https://stock.tuchong.com/image?imageId=465170289977458859.

Tuchong. (2020m). *Computer vision in intelligent city on auto-driving*. Retrieved June 8, 2020, from https://stock.tuchong.com/image?imageId=259828272896147689.

Wertheimer, M., et al. (2012). *On perceived motion and figural organization* (The MIT Press). MIT Press.

Wöhler, C. (2012). *3D computer vision: Efficient methods and applications* (X.media.publishing). Springer.

Wymann, O., et al. (2002). *The matrix and philosophy: Welcome to the desert of the real* (Popular Culture and Philosophy). Open Court.

6

Natural Language Processing

(Consider this scenario: A late evening around 10 pm, Prof Chang in his office using office IP phone to call IT support service)

Janet: Good evening, Prof Chang, I am Sam from IT support center. How can I help you?

Prof. Chang: Good evening Sam, good to hear that you are here at this hour. My PC just hung up and no response. Please help.

Janet: I see. Maybe you can switch off your PC and power it on again after 1 minute. See whether it works.

Prof. Chang: Okay, I do it now. (After 1 minute)

Janet: Is it okay now?

Prof. Chang: It came up with a blue screen and have several choices to press different functions. What should I do?

Janet: Good, it is the emergency system support screen. Please press F4 to continue the system bootup sequence and we can remote control your PC to fix for good.

Prof. Chang: Okay. I just pressed the F4 button. PC is now bootup and start remote connection to IT support center.

Janet: Yes. I can see it now. The problem is not too serious. Please give me 5 minutes to fix it. (After 4 minutes ....) Prof. Chang, you PC is okay now. Please have a look.

Prof. Chang: Excellent work! Everything is okay now. So many thanks, Sam.

Janet: You are welcome. Have a nice evening.

Prof. Chang: You too, bye!

Abstract This chapter begins with the introduction of human language and intelligence. We also introduce the six linguistics levels in human language.

© The Editor(s) (if applicable) and The Author(s), under exclusive license to Springer Nature Singapore Pte Ltd. 2020
R. S. T. Lee, *Artificial Intelligence in Daily Life*,
https://doi.org/10.1007/978-981-15-7695-9_6

Next, we study NLP main components that include natural language understanding (NLU), speech recognition, syntactic analysis, semantic analysis, pragmatic analysis, and speech synthesis followed by major NLP applications which include machine translation (MT), information extraction (IE), information retrieval (IR), sentiment analysis, question and answering (Q&A) chatbots.

Is Janet a human system support? Maybe. Can she be a service robot? The answer is definitely *yes*.

At the time the author wrote the book on intelligent agent over a decade ago (Lee 2006a, b), *Natural Language Processing (NLP)* technology was only a research topic. But today, NLP for customer service become so popular in major corporations that we won't even know the customer service agent is a robot or a human.

How can we do it?

In this chapter, we will introduce this fascinating technology by studying the main components of NLP. We will soon find out that NLP technology is closely related to different disciplines that include linguistics, statistical engineering, machine learning, data mining, human voice processing, etc. We will also be surprised by AI scientists' and NLP engineers' genius and efforts during the past two decades to turn this important research topic into commercial products that can be a great help in many applications to our daily activities.

Let us begin our journey on natural language processing with *human language and intelligence*.

6.1   Human Language and Intelligence

How we behave define who we are. The author thinks the clause is candid. Since we cannot know what people are thinking about, the only cue to evaluate or form the opinion of a person whether is good or bad, genius or dumb can only be achieved by observing their behaviors. The most direct way is to observe what he/she said by simple dialogues. This was why Sir Alan Turing; the father of AI devised the famous *Turing Test* (Shieber 2004) in the 1950s as a way to judge whether a machine has intelligence (Fig. 6.1).

Fig. 6.1 The Turing test

In terms of AI perspective, the core technology of *Turing Test* is in fact NLP (Eisenstein 2019), the technology to recognize and understand human language questions, and how to respond also in human language to the judge. It is the ultimate challenge of NLP.

Human language is a crucial component in civilization and one of the most fundamental aspects of our behaviors. In a general sense, it can be categorized into two main aspects: written and oral languages. For written language, the main function is to store and pass our knowledge to others of this and future generations. For oral language (spoken language), the core function is to act as a medium for communication behind a person to another in our daily activities. Language study is exceedingly important that different disciplines have their own focuses and interpretation. Each discipline comes with its own set of language-related problems to tackle and a set of solutions to address those problems. Table 6.1 shows a summary of these language disciplines and solutions to related problems (Eisenstein 2019; Bender 2013).

6.2 Levels of Linguistic in Human Language

Levels of linguistic (Hausser 2014) refer to the functional analysis of any human language that include both written and spoken languages. In terms of linguistic analysis, there are six levels of linguistics classified into three main categories. The two basic linguistic levels of sound include: *phonetics*

Table 6.1 Various disciplines related to NLP

Discipline	Problems to tackle with	Solutions and tools
Philosophy	What is meaning and knowledge?	Ontology and epistemology
	How do words and sentences acquire meaning?	Natural language argumentation using intuition
	How can we relate ideas and concept into words and meanings	Mathematical models such as logic theory and model theory
Psychology	How can we identify the structure of sentences?	Psychological experiments to measure the performance
	How the meaning of words can be identified? When does understanding take place?	Statistical analysis of observation
Linguistics	How to form phrases and sentences with words?	Mathematical model of language structure
	How can we representee meaning of a sentence?	Logical model for the representation of language structure and patterns
Computational linguists and NLP	How to model different type of human languages?	Agent ontology and ontological tree modeling
	How to model knowledge and meaning?	NLP techniques discussed in this chapter
	How to use human language for human-machine direct communication?	

and *phonology* sound levels. These two levels related to the sounds of spoken language.

Phonetics is about the physical aspect of sounds; it studies the production and perception of sounds called *phones*. *Phonetics* deals with the production of speech sounds by human, often without prior knowledge of the spoken language.

Phonology is about the abstract aspect of sounds and studies *phonemes*. *Phonology* is about establishing what are the phonemes of a given language, i.e. those sounds can deliver a different meaning between two words.

For instance, the vowels in the English words *cool*, *whose*, and *moon* are all similar but slightly different. The different variants depend on the different contexts in which they occur.

Intermediate linguistic levels of structure refer to the basic language which include two levels: *morphology* and *syntax levels* of language structure.

Morphology is the level of *forms* and *words*. It normally comprehends by grammar (along with syntax). The term *morphology* refers to minimal forms in language analysis, comprised of sounds to construct words that have either a grammatical or a lexical function. Lexicology is concerned with the study of the lexicon from a formal point of view and is thus linked closely to *(derivational) morphology*.

Syntax is the level of clauses and sentences. It is concerned with words meanings in combination with each other to form phrases or sentences. It involves particularly different meaning by changes in word order, addition or subtraction from sentences or changes in sentence forms. Furthermore, it deals with the relatedness of different sentence types and ambiguous sentences analysis.

Advanced linguistic levels of meaning refer to the actual meaning of the language, which include two levels: *semantic* and *pragmatic* levels of language meaning.

Semantics is the meaning area. It is thought that semantics is covered by areas of morphology and syntax. Thus, for every language level there exists lexical, grammatical, sentence, and utterance meaning.

Pragmatics is the use of language in specific situations. The meaning of sentences does not have to be identical to an abstract form in practical use. For utterance meaning, the area of pragmatics relies strongly for its analyses on the notion of speech act which concerned with actual language performance. This involves the notion of proposition—about the content of a sentence with its intent and effect of an utterance (Fig. 6.2).

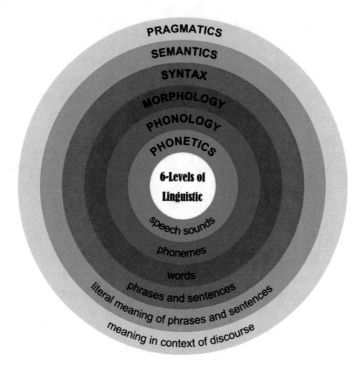

Fig. 6.2 Levels of linguistic in human languages

6.3   Ambiguity in Human Language

Ambiguity and *uncertainty* in *language ambiguity* used in natural language processing refer to the ability of being understood in more than a single way. Natural language is ambiguous inducing NLP with subsequent types of ambiguities.

Lexical Ambiguity

The ambiguity of a single word is called lexical ambiguity. For example, consider the word *silver* as a noun of metal, an adjective of silver colored, or a verb of process of silvering.

Syntactic Ambiguity

This kind of ambiguity occurs when a sentence is parsed in different ways. For example, the sentence: *The man saw a girl with telescope.* It is ambiguous whether the man saw the girl carrying a telescope or he saw her through his telescope.

Semantic Ambiguity

This kind of ambiguity occurs when the meaning of words can be misinterpreted. In other words, semantic ambiguity occurred when a sentence contains an ambiguous word or phrase. For example, the sentence: *The car hit the dog while it was moving* is having semantic ambiguity because the interpretations can be: *The car, while moving, hit the dog* and *the car hit the dog while the dog was moving*.

Pragmatic Ambiguity

Pragmatic ambiguity refers to a situation where the context of a phrase gives multiple interpretations. In simple words, we can say that pragmatic ambiguity arises when the statement is not specific. For example, Marie says: *I'll go to river bank this morning*. The meaning of river bank normally refers to probably the bank of the river. But if she says: *I'll go to riverbank this morning to take some cash*. In that case, the second part of the clause *to take some cash* gives us more clue about what she refers to riverbank maybe is the name of a bank. Figure 6.3 shows a typical example of ambiguity in a language in high level of pragmatic meaning.

6.4   A Brief History of NLP

First Stage—Machine Translation on NLP (Before 1960s)

The history of NLP can be traced back to the seventeenth century when philosophers such as polymath Gottfried Wilhelm Leibniz (1646–1716) and philosopher, mathematician, and scientist René Descartes (1596–1650) proposed to use codes to relate words between different languages. Although the proposals remained only theoretical then, they laid the ground for language translation machine development (Santilal 2020).

The first invention patent related to translation machine was applied by inventor and engineer Georges Artsrouni proposal in 1933. However, the

Fig. 6.3 What do you mean by *"bank"*

history of NLP *officially* began in the 1950s with Sir Alan Turing for his famous article publication *Computing Machinery and Intelligence* (Turing 1936, 1950) and the proposal of *Turing Test* to explore machine intelligence using NLP as judging criteria. At the time, NLP was mainly focused on intelligent machine R&D for language translation—*machine translation.*

The first international conference on *machine translation* (MT) was held in 1952 and the second one in 1956. NLP remained focused only on machine translation were used mainly on simple *rule-based* methods and statistical techniques. In 1954, *Georgetown-IBM experiment* involved fully automatic translation of more than 60 Russian sentences into English. The inventors at that time were overoptimistic to claim that the entire machine translation problem would be completed solved within 3–5 years. However, real progress took longer than expected. It was not until the *Noam Chomsky's Syntactic Structures* helped to revolutionize linguistics with universal grammar in 1957. The work presented in *Teddington International Conference on Machine Translation of Languages and Applied Language Analysis* reached its climax in 1961. After the release of the ALPAC report in 1966 revealed that the 10-year-long research had failed to fulfill its original expectation of machine translation in result of funding for all related research and projects were reduced substantially.

Second Stage—Early AI on NLP (Late 1960s–1970s)

When AI became popular at this period of time, major NLP development was focused on how AI can be applied to knowledge exploration, so-called *ontology* (Climiano et al. 2014), and its role on construction and manipulation of meaning representations. A typical example includes *BASEBALL* system development in the late 1960s, a question-answering AI-based expert system. However, the input to this system was restricted and involved simple language processing.

A more advanced NLP system was proposed by cognitive scientist Prof. Marvin Minsky (1927–2016) in 1968—a major founder of AI. Compared with *BASEBALL Q&A system*, this NLP system employed AI-based inference engine and knowledge base for Q&As interpretation.

Augmented Transition Network (ATN) was introduced to represent natural language input by distinguished software engineer Prof. William A. Woods in 1970. During that time, many programmers also began to write conceptual ontologies which structured real-world information into computer-understandable data. But because of these high expectations on AI and expert systems were not truly realized, both US Government and commercial sectors withdraw further research funding led to the *first winter of AI (1974–1980)*.

Third Stage—Grammatico-Logical on NLP (1970s–1980s)

This phase can be described as the *grammatico-logical phase*. Due to the failure of building practical system at the last phase, researchers moved toward the use of logic for knowledge representation and reasoning in AI.

The grammatico-logical approach provided succor and powerful general-purpose sentence processors like *SRI's core language engine* and *discourse representation theory*, which offered a means of tackling more extended discourse toward the end of decade. In this phase, practical resources and tools like parsers, e.g. Alvey natural language tools along with more operational and commercial systems for database query.

Although NLP R&D were bounded by the computational capacity of computer systems in this period, the work on lexicon provided a solid foundation and direction of grammatico-logical approach for future development.

Fourth Stage—AI and Machine Learning (1980s–2000s)

Most natural language processing systems were based on complex sets of hand-written rules until the 1980s. The rebirth of AI coined by physicist and scientist Emeritus Prof. John Hopfield success on his ground-breaking Hopfield Network in machine learning was revolutionary to introduce machine learning algorithms for language processing.

Thanks for the improvement on computer technology in terms of computational capacity and memory storage together with the dominance of Chomskyan theories of linguistics, whose theoretical underpinnings discouraged the sort of corpus linguistics machine learning approach to language processing. This stage was also known as NLP *lexical & corpus*. It had a lexicalized approach to grammar appeared in the late 1980s and became influential. Watson, a question-answering computer system capable of answering questions posed in natural language was developed by a research team of IBM DeepQA project led by scientist Dr. David Ferrucci in 2006.

Fifth Stage—AI, Big Data, and Deep Networks (2010–Now)

This period considers as years of *AI, Big Data, and Deep Networks*. With the advance of cloud computing technology, massive knowledge representation, data mining, and knowledge discovery favor AI-based NLP development, particularly in agent ontology R&D is the core knowledge center in every NLP system.

Representation learning uses agent ontology and deep neural network style machine learning methods became widespread in natural language processing in the 2010s. For example, *neural machine translation (NMT)* is a new type of deep network-based NLP system emphasizes on deep learning based approaches to machine translation directly learn sequence-to-sequence transformations, instead of using traditional *statistical machine translation (SMT)* (Koehn 2009). The technological maturity of voice recognition and human voice synthesis also favors NLP-based AI applications development such as AI chatbots, customer service robots, etc.

6.5   Natural Language Processing and AI

What is Natural Language Processing (NLP)?

NLP can be defined as human language automatic (or semiautomatic) processing (Eisenstein 2019). In some sense, the term NLP is sometimes narrowly used often excluding information retrieval and even machine translation. Many computer scientists consider NLP as *computational linguistics*. It is rather true in terms of computer science, NLP can be considered as a kind of *computer modeling* or *computerization* of linguistics, alike the term *computational finance* refers to computational modeling of finance theory. NLP is a multidiscipline topic that involves extensive knowledge and basic concepts on linguistics and logic theory in theoretical mathematics. At present, NLP research covers cognitive science, psychology, and even philosophy in terms of epistemology and ontology.

NLP and AI

NLP is a field of AI in which computers analyze, understand, and derive meaning from human language in a smart and useful way. By utilizing NLP, AI developers can organize and structure knowledge to perform tasks such as automatic summarization, translation, named entity recognition, relationship extraction, sentiment analysis, speech recognition, and topic segmentation. NLP is used to analyze text, allow machines to understand how human speak and respond.

This human–computer interaction enables real-world applications like automatic text summarization, sentiment analysis, topic extraction, named entity recognition, parts-of-speech tagging, relationship extraction, stemming, and more. NLP is frequently used for text mining, machine translation, and automated question-answering. With the rapid growth of AI and computer technology, current NLP research and implementation also involve *AI-based machine learning, data mining, deep learning, and agent ontology* (Fig. 6.4).

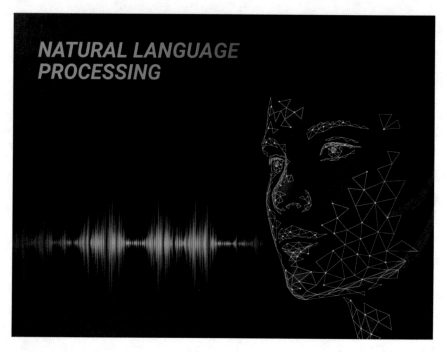

Fig. 6.4 NLP and AI

6.6   Main Components of NLP

NLP consist of three main components.

Natural Language Understanding (NLU)

NLU corresponds to all basic functions and operations of NLP from human language voice recognition to three levels analysis of understanding the meaning of spoken language: *syntax, semantic, and pragmatic analysis*. Traditional NLP with a restricted domain of application is the only focus of this component (Eisentsten 2019).

Knowledge Acquisition and Inferencing (KAI)

Once the spoken language is *clearly understood* from NLU, KAI focuses on proper response and answer generation. In terms of machine learning and AI, it is a kind of knowledge acquisition and inferencing problem. Traditionally, such a task is achieved by the rule-based system. That is an *if–then* kind of question and response which is frequently used in many expert systems. However, with the complexity of natural language and conversation, most rule-based systems failed to apply successfully. To solve this intrinsic problem, most KAI system will attempt to restrict the knowledge domain to certain area such as customer service knowledge for specific industry, e.g. insurance, IT. With the advancements in AI technology, a new technology on agent ontology has been implemented with certain success. We will study in detail about the Ontological-based Search Engine in the next chapter.

Natural Language Generation (NLG)

NLG involves reply, respond, and feedback generation in the human–machine conversation. This consists of the process of response formulation into texts and sentences with the target language, text-to-voice synthesis based on the target language to produce near-human voice response (Fig. 6.5).

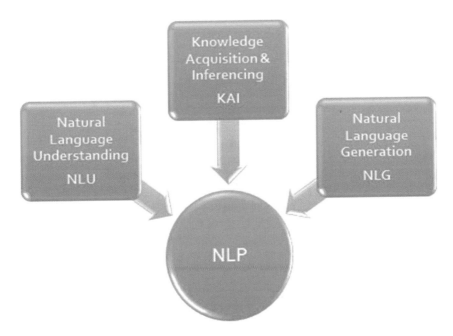

Fig. 6.5 NLP main components

6.7   Natural Language Understanding (NLU)

Natural language understanding (NLU) focuses on the *recognition and understanding* of the spoken language (Allen 1994). It consists of four main processes: speech recognition, syntax analysis, semantic analysis and pragmatic analysis. Figure 6.6 shows the systematic diagram of NLU.

Speech Recognition

It is the NLP first phase, which corresponds to phonetics implementation, phonology, and morphological processing of the spoken language mentioned

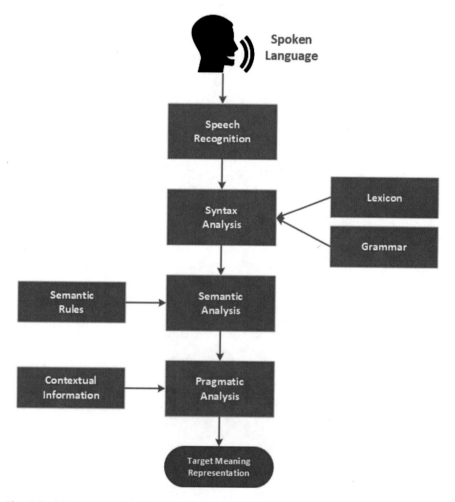

Fig. 6.6 NLU systematic diagram

in the linguistic model. The main purpose of this phase is to break chunks of spoken language input into sets of *tokens,* which correspond to paragraphs, sentences, and words. Current speech recognition applies frequency spectrogram technology to extract different frequencies of spoken sounds for speech recognition. For example, a word like *uncertain* can be broken into two sub-word tokens as *un-certain.*

Syntax Analysis

It is the NLP second phase, which corresponds directly to the first level for the structural meaning analysis of spoken sentence(s). The purpose of this phase is twofold: (1) to check that a sentence is well formed and (2) to break the spoken sentence(s) into a syntactic structure that can reflect the syntactic relationships between different words. For example, the sentence: *The apple goes to the girl* would be rejected by syntax analyzer or parser.

Semantic Analysis

It is the NLP third phase which corresponds directly to the second level for semantic meaning analysis of the spoken sentence(s). The purpose of this phase is to extract the exact meaning, or one can say the meaning defined by the dictionary extracted from the text. In order words, the extracted text is checked for its meaningfulness (or rejected with meaningless). For example, the semantic analyzer would reject a sentence like *hot snowflakes.*

Pragmatic Analysis

It is the NLP fourth phase and the most difficult level of meaning analysis of the spoken sentence(s). *Pragmatic analysis* deals with *outside world knowledge*, which means knowledge is external to the spoken sentence(s). Pragmatics analysis focuses on what was described is reinterpreted by what it actually meant, deriving various language aspects from real-world knowledge. For example, the sentence: *Will you crack the door? Is getting hot.* Semantically, the word *crack* would mean to break, but pragmatically we know that the speaker means to slightly open the door to let in some air.

6.8   Speech Recognition

6.8.1   Speech Recognition: The Basics

Speech recognition (Li et al. 2015)—technically known as the *voice-to-text* process is a complex process which consists of four basics steps: (1) *Voice capture/recording*; (2) *Analog-to-Digital Conversion (ADC)*; (3) Frequency spectrogram generation, and (4) Phonemes/words generation as shown in Fig. 6.7.

In voice capture/recording step, spoken voice is captured or recorded voice communication devices such as PC's microphones or the user's mobile phone, via internet or mobile network to NLP system such as technical support of IT computer.

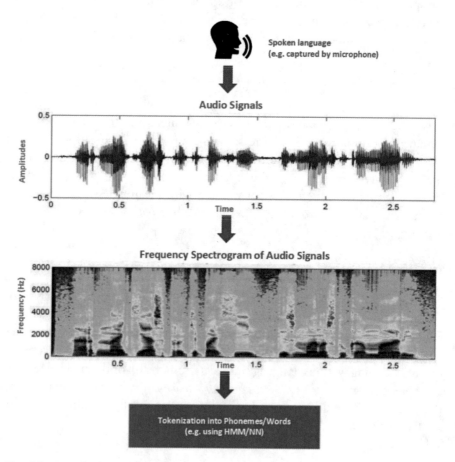

Fig. 6.7 Speech recognition basic operation

In the *Analog-to-Digital Converter (ADC)* step, the system translates these analog sound wave into digital data so that it can be processed by the speech recognition system. To achieve this, the system samples or digitizes the sound wave by taking measurements of sound waves at different time intervals.

In the frequency spectrogram generation step, the system filters background noises to improve sound quality, and then separate the sound waves into different bands of frequency to generate the so-called *Frequency Spectrogram*. It also normalizes the sound or adjusts it to a constant volume level. As human do not always speak at the same speed, so the sound must be adjusted to match with the speed of template sound samples already stored in the system's memory.

In *phonemes/words tokenization* step, the processed signal is divided into small segments as short as a few hundredths of a second, or even thousandths in the case of plosive consonant sounds—consonant stops produced by obstructing airflow in the vocal tract—like *p* or *t*. The system then matches these segments to known phonemes with the appropriate language. A phoneme is the smallest element of a language—a representation of sounds we make and put together to form meaningful expressions. There are approximately 40 phonemes in the English language (different linguists have different opinions on the exact number), while other languages have more or fewer phonemes. The most widely commercially used voice-to-text tokenization technology is based on *hidden Markov model (HMM), deep neural networks* (DNN) are also used nowadays. Next, let us take a look at how HMM works.

6.8.2 Hidden Markov Model (HMM)

Hidden Markov model (HMM) is a powerful statistical model based on the *Markov process* concept that outputs a sequence of symbols or quantities. HMM is used for the voice signal tokenization because it can be regarded as a piecewise stationary signal. Human voice can basically be modeled as a stationary process within a short timescale such as 5–10 ms. Speech can be thought of a *Markov model* for many stochastic purposes (Gales and Young 2008).

Another reason why HMM is popular because it is: (1) reliable; (2) easy to implement technically; (3) the *Markov chains* can be trained automatically and computationally feasible to use.

HMM is a system where a variable can switch (with varying probabilities) between several states, generating one of several possible output symbols with each switch (also with varying probabilities). The sets of possible states and unique symbols may be large, but finite and known.

The basic process in HMM in speech recognition are.

(1) *Inference:* given a specific sequence of output symbols, compute the probabilities of one or more candidate state-switch sequences.
(2) *Pattern matching:* find the state-switch sequence most likely to have generated a specific output-symbol sequence.
(3) *Training:* given examples of output-symbol sequence (training) data, compute the state-switch/output probabilities (i.e. system internals) that fit this data best (Fig. 6.8).

Figure 6.8 shows a typical HMM model of spoken voice tokenization at phonemes and words level.

As shown, a comprehensive lexicon for the targeted language is required to accomplish the speech recognition process. We will study about it in the coming section.

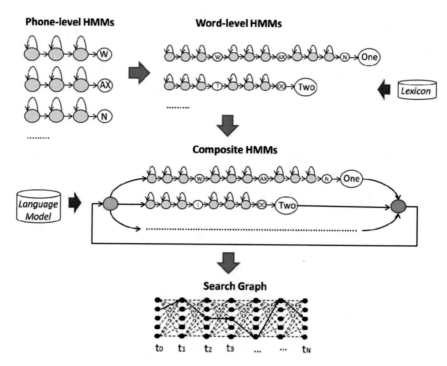

Fig. 6.8 Hidden Markov model on speech recognition

6.9   Syntactic Analysis

6.9.1   Parsing

Syntax analysis (Sportier et al. 2013) is also known as *parsing*. Parsing is the process of determining whether a string of tokens can be generated by the grammar. It is performed by syntax analyzer which can also be termed as a parser. Figure 6.9 shows the basic parsing mechanism in syntactic analysis.

As shown, tokenized input texts (words or phonemes) generated by the speech recognition system are fed into the *lexical analyzer* and cross-checked with the lexicon database such as *WordNet* to examine for correct syntax and grammar.

After that, it passes into the parser to establish a data structure generally in the form of a *parse tree* or other syntax structures.

The main roles of the parse include to.

- check for any syntax or grammatic error,
- recover from commonly occurring error so that the processing of the remaining program can be continued,
- construct and modify the parse tree,
- construct and modify the symbol table, and
- produce intermediate representations for information retrieval.

It may be defined as a software component designed for taking input data (text) and giving input structural representation after checking for correct syntax as per formal grammar. It also builds a general data structure in the form of a parse tree or abstract syntax tree or other hierarchical structure.

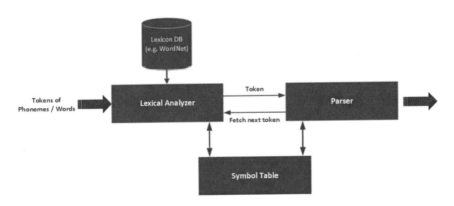

Fig. 6.9 Levels of linguistic in human languages

6.9.2   Parse Tree

Grammar in Parser

Grammar is essential and important to describe the syntactic structure of well-formed languages (and even programming languages). In the literary sense, it defines conversation syntactical rules for different languages. Linguistics have attempted to define grammar since the inception of natural languages like English, Chinese, Japanese, Hindi, etc. The theory of formal languages is also applicable in the fields of computer science mainly in programming languages and data structure. Even for a computer language such as *Java,* the precise grammar rules state how functions are made from lists and statements.

What is a Parse Tree?

Parse Tree is defined as the graphical depiction of a derivation. The start symbol of derivation serves as the root of the parse tree—the *sentence(s)* root node.

The other nodes include all constitutional components in a sentence that include: *Noun_Phrase (NP), Verb_Phrase (VP), Determiner (D), Verb (V), Noun (N), Proper_Noun (PN), Adjective (AJ), Adverb (AV),* etc. For each parse tree, the leaf nodes are terminals and interior nodes are nonterminals. A property of parse tree is that in-order traversal will produce the original input text sentence.

So, the main function of parsing is: Given a sentence with a grammar, the parser will check the sentence whether it is correct according to grammar and if so, returns a parse tree representing the sentence structure. Figure 6.10 shows an example of parse tree for the sentence: *This diagram is illustrating the parsing tree.*

6.10   Semantic Analysis

What is Semantic Analysis?

Semantic analysis (Goddard 1998) is to extract the actual meaning of the text or may say the *dictionary meaning.* Anyway, the main task of semantic analysis is to examine the *meaningfulness* of the text. One may wonder: Lexical analysis in syntactic analysis in some sense already examined the meanings of words, so what is the difference between lexical and semantic analysis?

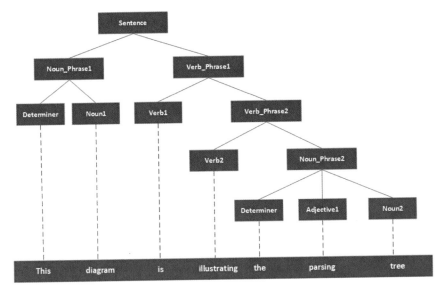

Fig. 6.10 An example of *parse tree* for the sentence: *This diagram is illustrating the parsing tree*

The truth is: *Lexical analysis* only looks into the meaning of individual words from lexicon, while *semantic analysis* extracts the overall meaning of the text that most likely involves the combination of more than one or several words token in order to extract the actual meaning of the complete text message.

For example, the sentence: *Einstein is great scientist* means *Albert Einstein, the one who proposed General Relativity is a great scientist*; or another scientist called *Einstein* is a great scientist? To solve this problem, we need to know more about the sentence to provide more cues which *Einstein* refers to, in order to investigate the actual meaning of the sentence. That is the reason why we need semantic analysis.

Semantic Network

A *semantic network* (Aggarwal 2018) is a *knowledgebase* that represents semantic relations between concepts in a network. This is often used as a form of knowledge representation.

In short, a semantic network is a directed or undirected graph consisting of vertices, which represent concepts, and edges that represent semantic relations between concepts, mapping, or connecting semantic fields. In general, most semantic networks are cognitively based. They also consist of arcs and

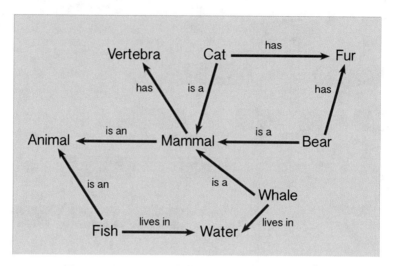

Fig. 6.11 An example of *Semantic Network* for meanings and concepts related to *Mammal*

nodes, which can be organized into a taxonomic hierarchy. Figure 6.11 shows a typical concept example of a semantic network for the word (concept) *mammal*. As shown, we can extract several useful knowledges from a simple semantic network.

To use the semantic network, simply start with the node corresponds to the *word* we interest, forward an arrow of the graph to find a path. Each path corresponds to a specific knowledge related to the concept we interest. For instance:

- *Mammal* is an animal.
- *Bear* is a mammal with vertebra.
- *Whale* is a mammal, lives in water.
- *Cat* is a mammal has fur.

6.11 Pragmatic Analysis

What is Pragmatic Analysis?

Pragmatic analysis (Eisenstein 2019; Ibileye 2018) is the last phase of linguistic analysis. It is part of the process of extracting information from text. Specifically, it is the portion that focuses on taking a text structure set and reasoning out the actual meaning.

Unlike semantics which examines meaning that is conventional or *coded* in a given language, pragmatic analysis studies how the transmission of meaning depends not only on structural and linguistic knowledge (such as grammar and lexicon) of the speaker but also on the context of the utterance, that might involve any pre-existing knowledge and the inferred intent of the speaker. The most famous example is the clause: *Raining cats and dogs.* We all know it means raining heavily. But how can we draw this knowledge. As such concept of knowledge is totally unrelated to the syntax, semantic, or even the semantic networks of either: cat, dog, or rain—this is the task for *pragmatic analysis*. In other words, the main purpose of pragmatic analysis is trying to extract the *true knowledge (meaning)* of spoken speech, which may be or may not be directly reflected by its semantic meaning. Due to high complexity and ambiguity to extract the *embedded meaning* of spoken language, not only related to the spoken message, but also other related knowledge and concepts outside the topic context and knowledge domain.

Pragmatic analysis is believed to be one of the most difficult topics in linguistic and AI in terms of NLP implementation with knowledgebase and search engine. It is different from voice recognition; syntactic and semantic analysis are technically mature enough as widely adopted methods and technology. Pragmatic analysis is still in the R&D stage without any dominant technology and solution.

Latest research of pragmatic analysis includes R&D of an emerging AI technology—Agent Ontology, which focuses on the ultimate problem of how human knowledge is generated, stored, and retrieval. More importantly, different concepts and ideas are related together to retrieve actual and embedded meaning, which will be studied in the next chapter—*Ontological-based Search Engine.* Figure 6.12 shows a snapshot of an example ontology graph.

6.12 Speech Synthesis

Text-To-Speech Synthesis

Speech synthesis (Taylor 2009) is an artificial simulation of human speech by a computer or other device mostly used for translating text information into audio information as a counterpart of voice recognition. It is also known as TTS (*text-to-speech*) technology.

Further, speech synthesis is an assistive technology used by vision-impaired individuals to read text contents. It is alike speech recognition as a mature

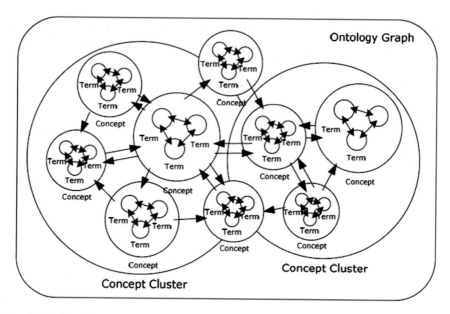

Fig. 6.12 Ontology graph conceptual diagram (Lim et al. 2011)

technology widely used in many related applications such as voice-enabled services and mobile applications of our daily activities.

Speech Synthesis System

A typical speech synthesis system consists of three main modules: *text analyzer; linguistic analyzer,* and *waveform generator* as shown in Fig. 6.13. The main function of the *text analyzer* is to convert response text generated,

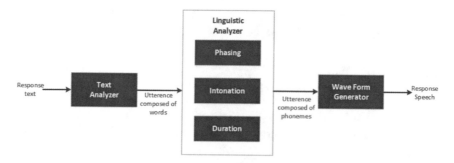

Fig. 6.13 Speech synthesis systematic diagram

says, from ontology-based knowledgebase into words tokens—*tokenization process*. After that, it passes to the *linguistic analyzer* for further processing.

In the *linguistic analyzer*, it assigns phonetic transcriptions to each word, divides and marks the text into prosodic units, like phrases, clauses, and sentences. The process of assigning phonetic transcriptions to words is called *text-to-phoneme* or *grapheme-to-phoneme* conversion. Phonetic transcriptions and prosody information jointly make up the symbolic linguistic representation. The processed phonemes then pass to the *waveform generator*.

The *waveform generator*, commonly known as *speech synthesizer* converts the symbolic linguistic representation into sound. In most systems, this part includes the computation of target prosody (pitch contour, phoneme durations), which is then imposed on output speech in the form of human voices.

Corpus

In linguistics, *corpus* (Szudarski 2017) is a large and structured set of machine-readable texts produced in a natural communicative setting. Technically speaking, a corpus can be derived in different ways like text that was originally electronic transcripts of spoken language, optical character recognition, etc. Language is infinite but a corpus must be finite in size, we need to sample and proportionally include a wide range of text types to ensure a good corpus design. Another important element of corpus design is its size. How large a corpus should be? There is no specific answer to this question. Corpus size depends upon the purpose of intension with other practical considerations as follows:

- Kind of query anticipated from users.
- The methodology used by users to study the data.
- Availability of the source of data.

Corpus size also increases with technology advancement. For example, the size of a *Brown and LOB* corpus is around 1 million words, while the frequently used *Bank of English* corpus today is over 650 million words.

TreeBank and ProBank Corpus

TreeBank corpus (Abeillé 2003; Linguistic Data Consortium 2020) is linguistically parsed text corpus that annotates syntactic or semantic sentence structure. Specialist in English language and linguistics Emeritus Prof. Geof-frey Leech (1936–2014) coined the term *treebank* as the most common way of representing grammatical analysis by means of a tree structure. *Semantic* and *syntactic treebanks* are the two most common types of treebanks in linguistics.

ProBank Corpus also known as *proposition bank* annotated with verbal propositions and their arguments was developed by computer scientist Prof. Martha Palmer et al. The corpus is a verb-oriented resource, annotations here are more closely related to the syntactic level. In NLP, *PropBank* project has played a significant role to assist in semantic role labeling. Figure 6.14 shows a sample snapshot of the parse tree construction using TreeBank Corpus.

WordNet and VerbNet

WordNet (Cognitive Science Laboratory 2009) and VerbNet (Bunt et al. 2013) are the two most frequently used corpus in NLP.

VerbNet (VN) is the hierarchical domain-independent and largest lexical resource present in English that incorporates both semantic as well as syntactic information about its contents. VN is a broad-coverage verb lexicon having mappings to other lexical resources such as *WordNet, Xtag, and*

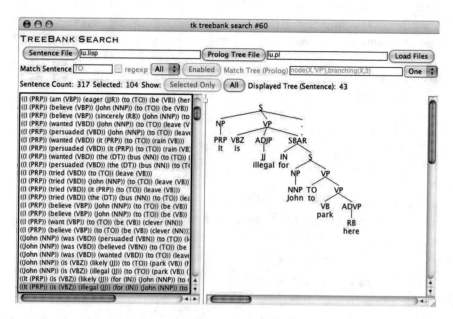

Fig. 6.14 Snapshot of parse tree using TreeBank Corpus

FrameNet. It is organized into verb classes by refinement and addition of subclasses to achieve syntactic and semantic coherence among class members.

WordNet created by Princeton University is a lexical database for the English language. WordNet is a lexical database of semantic relations between words in more than 200 languages and the most frequently used corpus in the world, both for commercial and academic research in NLP, knowledge-base, and agent ontology R&D. In WordNet, nouns, verbs, adjectives, and adverbs are grouped into sets of cognitive synonyms called *synsets*. All synsets are linked with the assistance of conceptual–semantic and lexical relations.

In terms of information systems, WordNet is used for various purposes like word-sense disambiguation, information retrieval, automatic text classification, and machine translation. One of the most important uses of WordNet is to find out the similarity among words. Due to its popularity, various functions and algorithms on syntactic and semantic analysis have been implemented in various NLP-related development platforms as functional libraries or packages that include C, C++ , Perl, ADW in Java, and NLTK in Python. Figure 6.15 shows a snapshot WordNet database from the perspective of adjective "*good*".

6.13 Applications of NLP

After over 20 years of R&D and actual implementation, NLP technology is now being used in many applications that are closely related to our daily activities. They include *machine translation* (MT), *information extraction* (IE), *information retrieval* (IR), *sentiment analysis, question & answering* (Q&A) *robots* (Systems), etc. as shown in Fig. 6.16.

Machine Translation (MT)

Machine translation (Scott 2018) is the earliest, most well-studied and one of the most important NLP applications. A major challenge in MT nowadays is twofold: (1) the *naturalness* (or *fluency*)—machine translation that is natural in the target language while preserving the exact meaning expressed by input; (2) the *adequacy*—the degree of MT to which output reflects the meaning of the source.

These two are often in conflict, especially when the source and target languages are not similar (e.g. translation between Chinese and English). Experienced human translators address this trade-off in an artistic way.

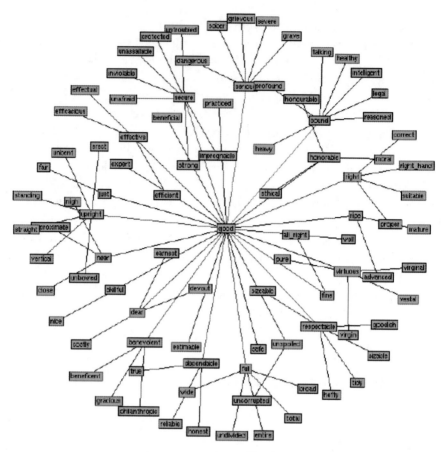

Fig. 6.15 Parse tree using TreeBank Corpus snapshot

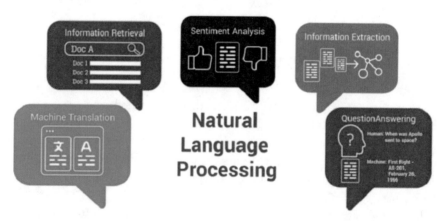

Fig. 6.16 NLP applications

The goal of nowadays MT is to apply various AI technologies such as deep learning to learn from experts to achieve human-quality translations.

Information Extraction (IE)

Information extraction (IE) (Hemdev 2011) is the task of extracting structured information automatically from unstructured and/or semi-structured machine-readable documents and other electronically represented sources. In most of the cases, this activity concerns processing human language texts by means of NLP.

Recent activities in multimedia document processing like automatic annotation and content extraction out of images, audio, video, and text documents could be regarded as information extraction. Due to the difficulty of the problem, many commercial IE applications are domain specific such as the focus of a specific discipline (e.g. law, environmental) or topic interest.

Information Retrieval (IR)

Information retrieval (IR) (Peters 2012) is a software program that deals with the organization, storage, retrieval, and evaluation of information from document repositories particularly textual information. The system assists users in finding the information they require but does not explicitly return answers to the questions. It informs the existence and location of documents that might consist of the required information. The documents that satisfy the user's requirement are called relevant documents. A user who needs information will have to formulate a request in the query form using NLP. Then the IR system will respond by retrieving the relevant output, in documents form about the required information. In fact, the main objective of the IR system is to develop a model to retrieve information from the repositories of documents.

A typical example of an IR system is a so-called ad hoc *retrieval problem*. Figure 6.17 shows the process flowchart of a typical information retrieval system using ad hoc retrieval method. In ad-hoc retrieval, the user must enter a query in natural language that describes the required information. Then the IR system will return the required documents related to the desired information. For example, suppose we search something from the Internet, and it gives some exact pages that are relevant per our requirements but there can contain some nonrelevant pages. This is due to the ad-hoc retrieval problem.

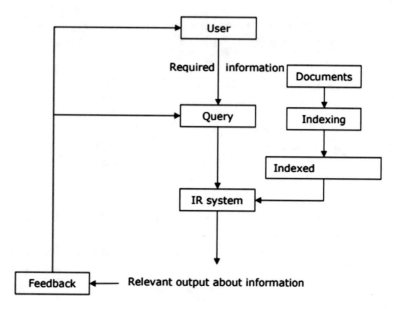

Fig. 6.17 Process flowchart of an information retrieval system

Sentiment Analysis

Sentiment analysis (Liu 2012) is a type of data mining that measures the inclination of people's opinions through NLP, which are used to extract and analyze subjective information from the Web, usually social media, and similar sources. The analyzed data quantifies the general public's sentiments or reactions toward certain products, people, or ideas and reveal the contextual polarity of the information.

Customer services use sentiment analysis, an NLP application to identify user's opinion and sentiment. It will enable companies to understand what their customers think about the products and services. Companies can also judge their overall reputation from customer posts with the assistance of sentiment analysis. In this way, we can say that beyond determining simple polarity, sentiment analysis understands sentiments in context allowing us to better understand what is behind the expressed opinion. Figure 6.18 shows a typical scenario of sentiment analysis using NLP technology.

As shown, with the integration of data mining technology and NLP, user's responses and comments on various topic of interest can be converted into machine-understandable concepts and ideas that can be classified effectively into different degrees of emotion and response, that can be used by companies to better understand and analyze customers' needs. For news agency and

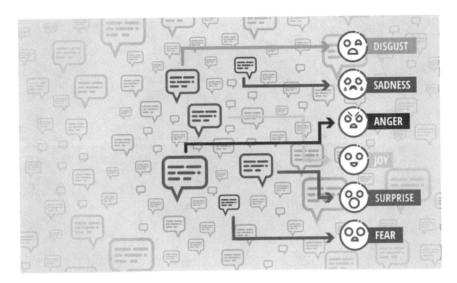

Fig. 6.18 NLP on *Sentiment Analysis*

public forum, sentiment analysis together with NLP technology can be used to data-mine public opinions and comments more effectively and objectively.

In fact, NLP-based sentiment analysis is widely used to major social media and forum to users' opinion and real-time responses to some ad-hoc events and incidences.

Question and Answering (Q&A) Robots (Systems)

Another main application of natural language processing (NLP) is question-answering robots, or so-called *chatbot* (Raj 2018). In general, Q&A systems is the *ultimate challenge* of NLP and AI which is the main theme of the Turing Test. It concerns with the building of the AI system and automatically answers questions raised by a human using our own languages. In other words, the Q&A not only need to recognize and understand human language, but also need to know the *actual* meaning from syntax, semantic up to pragmatic levels.

It also needs to response with a human voice, involves high-level knowledge-based and inferencing, together with human voice generation system. With the rapid growth in AI and NLP technology, Q&A robots and systems are widely used in many industries that include.

Fig. 6.19 Customer services robots using NLP technology (Tuchong 2020a)

- Technical support robots, e.g. provide IT basic technical support via internet and traditional hotlines.
- Customer service robots, e.g. promote products and after-sale support services.
- Language learning tutor, e.g. teach and train students in language centers.

Figure 6.19 shows a typical scenario of Q&A customer services robots.

6.14 Case Study: Language Learning Robots Using NLP

Companion robot is not a new thing in AI world. It dates back to the 1980s, where Japanese industries had already developed several famous companion robots to interact with human and provide limited NLP capability.

With the advanced AI, computing, and robotic technology, nowadays companion robots are capable to provide dynamic services, such as foreign language learning robots (LLR) to teach students how to speak foreign languages, i.e. English LLR teach Asian students to learn English, or Spanish LLR teach English-language-oriented students to learn Spanish in their daily activities.

Fig. 6.20 An English-language learning robot (Tuchong 2020b)

Suppose you are an English LLR system designer. Based on the NLP technology learnt in this chapter, together with various machine learning and data mining techniques learnt from previous chapters:

(1) What are the three basic machine learning techniques in AI?
(2) Discuss and explain how these machine learning techniques can be applied to English learning?
(3) Discuss and explain how to integrate these machine learning techniques, together with the NLP technology to implement an English Learning Robot?
(4) Many learning systems have different levels of challenges. Suppose you need to design this English LLR with 3 levels of challenges. What are these 3 levels of challenges?
(5) What kinds of AI and NLP methods do you use to implement these 3 levels of function? (Fig. 6.20)

6.15 Conclusion

In this chapter, we discuss an exceedingly challenging and important AI technology—*natural language processing (NLP)*. NLP is not a new topic. In fact, the *Turing Test* proposed by Sir Alan Turing in the 1950s was the focus on

NLP performance of the machine to determine the degree of intelligence. It was also one of the core components of *Generalized AI (GAI)* and robot design during the 1960–80s.

However, owing to AI over-expectation and computational capability limitation, NLP technology development was sluggish and mainly focused on statistical-based machine learning (ML) applications. With the advancements in computational speed and AI popularity, machine learning, deep network, big data and data mining, NLP technology and related applications had evolved rapidly in the past 20 years.

Today, various NLP-related technology such as human voice synthesis systems for car navigation, information retrieval (IR), information extraction (IE), NLP-based customer services robots, sentiment analysis in social media, machine translation apps, and systems with Q&A chatbots became part of our daily activities.

NLP is in fact not only a human voice related technology, but also closely related to how we acquire, learn, store, and manipulate our knowledge. The so-called *ultimate challenge* of AI–knowledgebase and agent ontology problem.

In the next chapter, we will study this challenging topic—Ontological-based Search Engine (OSE) (Fig. 6.21).

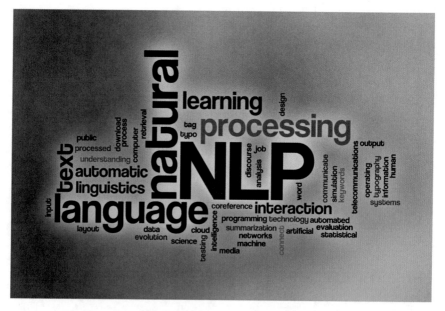

Fig. 6.21   NLP framework (Tuchong 2020c)

References

Abeillé, A. (Ed.). (2003). *Treebanks: Building and using parsed corpora* (Text, Speech and Language Technology Book 20). Springer.

Aggarwal, C. C. (2018) *Neural networks and deep learning: A textbook.* Springer.

Allen, J. (1994). *Natural language understanding* (2nd ed.). Pearson.

Bender, E. M. (2013). *Linguistic fundamentals for natural language processing: 100 essentials from morphology and syntax* (Synthesis Lectures on Human Language Technologies). Morgan & Claypool Publishers.

Bunt, H., et al. (2013) *Computing meaning: Volume 4* (Text, Speech and Language Technology Book 47). Springer.

Climiano, P., et al. (2014.) *Ontology-based interpretation of natural language* (Synthesis Lectures on Human Language Technologies). Morgan & Claypool Publishers.

Cognitive Science Laboratory. (2009). *WordNet 3* (largest English dictionary and thesaurus, Kindle edition). OSNOVA.

Eisenstein, J. (2019) *Introduction to natural language processing* (Adaptive Computation and Machine Learning series). The MIT Press.

Gales, M., & Young, S. (2008). *Application of hidden Markov models in speech recognition* (Foundations and Trends(r) in Signal Processing). Now Publishers Inc.

Goddard, C. (1998). *Semantic analysis: A practical introduction* (Oxford Textbooks in Linguistics). Oxford University Press.

Hausser, R. (2014). *Foundations of computational linguistics: Human-computer communication in natural language* (3rd ed.). Springer.

Hemdev, P. (2011). *Information extraction: A smart calendar application: Using NLP, computational linguistics, machine learning and information retrieval techniques.* VDM Verlag Dr. Müller.

Ibileye, G. (2018). *Discourse analysis and pragmatics: Issues in theory and practice.* Malthouse Press.

Koehn, P. (2009). *Statistical machine translation.* Cambridge University Press.

Lee, R. S. T. (2006a). *Fuzzy-neuro approach to agent applications: From the AI perspective to modern ontology.* New York, Berlin: Springer.

Lee, R. S. T. (2006b). Lee-Associator—A chaotic auto-associative network for progressive memory recalling. *Neural Networks, 19*(5), 644–666.

Li, J., et al. (2015). *Robust automatic speech recognition: A bridge to practical applications.* Academic Press.

Lim, E. H. Y., Liu, J. N. K., & Lee, R. S. T. (2011). *Knowledge seeker: Ontology modelling for information search and management: A compendium.* Berlin: Springer.

Linguistic Data Consortium. (2020). Retrieved May 8, 2020, from Official site: https://www.ldc.upenn.edu/.

Liu, B. (2012) *Sentiment analysis and opinion mining.* Morgan & Claypool Publishers.

Peters, C., et al. (2012). *Multilingual information retrieval: From research to practice.* Springer.

Raj, S. (2018). *Building Chatbots with Python: Using natural language processing and machine learning.* Apress.

Santilal, U. (2020). *Natural language processing: NLP & its history* (Kindle edition). Amazon.com.

Scott, B. (2018) *Translation, brains and the computer: A neurolinguistic solution to ambiguity and complexity in machine translation* (Machine Translation: Technologies and Applications Book 2). Springer.

Shieber, S. M. (Ed.). (2004). *The turing test: Verbal behavior as the hallmark of intelligence* (A Bradford Book). MIT Press.

Sportiche, D., et al. (2013). *An introduction to syntactic analysis and theory.* Wiley-Blackwell.

Szudarski, P. (2017). *Corpus linguistics for vocabulary* (Routledge Corpus Linguistics Guides). Routledge.

Taylor, P. (2009). *Text-to-speech synthesis.* Cambridge University Press.

Tuchong. (2020a). *Customer service robots using NLP technology.* Retrieved May 8, 2020, from https://stock.tuchong.com/image?imageId=473008785683513407.

Tuchong. (2020b). *An English language learning robot.* Retrieved June 10, 2020, from https://stock.tuchong.com/image?imageId=487771704973328499.

Tuchong. (2020c). *NLP framework.* Retrieved May 8, 2020, from https://stock.tuchong.com/image?imageId=261988297848651796.

Turing, A. (1936). On computable numbers, with an application to the Entscheidungs-problem. *Proceedings of the London Mathematical Society,* Series 2, *42,* 230–26

Turing, A. (1950). Computing machinery and intelligence. *Mind, LIX, 236,* 433–460.

7

Ontological-Based Search Engine

Without computers, in the 17th century, we could classify the entire animal kingdom...
there was this idea of the speciation, right? And now, all a search engine is essentially
the mathematical speciation of ideas - and these things really derive from the way that
language is used and the way words relate.
Mr. Joshua Cohen (Writer, born 1980)

Abstract Ontology is a fundamental form of knowledge representation of the real world. From a computer science perspective, ontology defines a set of representational primitives to model a domain of knowledge. A well-constructed ontology can help to develop knowledge-based information search and management system such as search engine and content management system in a more effective way. We explore this challenging AI technology and study how such innovative technology can be used in our daily activities. The first part of the chapter reviews traditional search engines, their main components, basic architecture, and major shortcomings in terms of system and user perspectives. The second part of the chapter examines ontological-based search engine (OSE) with the basic concepts of knowledge and ontology such as ontology engineering, semantic web; how to use ontology graph (OG) to represent concepts and ideas followed by system architecture. It covers several major OSE applications include content management system, news retrieval and ontological search engine, and web ontology learning systems.

World Wide Web (the *Web*) provides a new era of information sharing and worldwide communication since the 1990s. There were over millions of new websites created on a monthly basis with overflooding of data almost without bound in the past 30 years. We need a software tool to help us search over the web and extract useful information over this unbounded territory, such a tool is called *search engine*. Commonly used search engines include Google, Bing, and Baidu are mainly based on keywords as the searching method which always resulted in related search results. A new kind of search engine is strongly needed. With the advancements in AI technology, an ontological-based search engine which is based on the literal meaning of the whole web document to provide search results seems to provide a more efficient and sensible solution.

Ontology is a fundamental form of knowledge representation of the real world. From a computer science perspective, *ontology* defines a set of representational primitives to model a domain of knowledge. A well-constructed ontology can help to develop knowledge-based information search and management system, such as a search engine and content management system in a more effective way.

We will explore this challenging AI technology in this chapter and study how such innovative technology can be used in our daily activities. The first part of the chapter will review traditional search engines, their main components, basic architecture, and major shortcomings in terms of system and user perspectives. The second part of the chapter will examine *ontological-based search engine (OSE)* with the basic concepts of *knowledge* and *ontology* such as ontology engineering, semantic web; how to use ontology graph (OG) to represent concepts and ideas followed by the system architecture. It will cover several major OSE applications that include content management system, news retrieval and ontological search engine, and web ontology learning systems.

7.1   World Wide Web and Search Engine

Digital information and worldwide knowledge are stored and managed in specific systems prior to the 1990s. Information sharing between different individuals and organizations are limited to propriety networks and systems. The rise of Internet technology, a new idea so-called *World Wide Web* (the *web*) was proposed by computer scientist Sir Tim Berners-Lee in the 1990s. Web (with *web servers*) offers a completely new way of *free information* (mainly on text and images at the beginning) between any computer systems worldwide with *browser*—a simple but powerful software for us to *surf* any websites, read and enjoy the sharing of information and knowledge (Berners-Lee 2000; Levene 2011).

There had been billions of web servers established over the past 25 years ranging from personal blogs to sophisticated corporate websites and government repository archives for access via this extraordinary invention. *Netcraft*, an Internet monitoring company that has tracked web growth since 1995, reported that there were 215,675,903 websites with domain names and content in 2009, compared to merely 19,732 websites in August 1995. There are over 1.5 billion websites on the World Wide Web as of April 20, 2020 with less than 200 million are active. However, the numbers of websites are still growing at an astonishing rate on a daily basis. Figure 7.1 shows worldwide website growth in the past 20 years.

So, the problem is: *Now can we look for information we want in this ocean of information source—search engine?* A search engine (or formally known as *web search engine*) is a GUI-based (*graphical user interface*) software system that is designed to carry out web search over the Internet. Technically speaking, the main function of search engine is to search the *web* in a systematic way for specific information by the user in a textual way, so-called *search query*. The search results are presented in a line of results generally, often referred to as

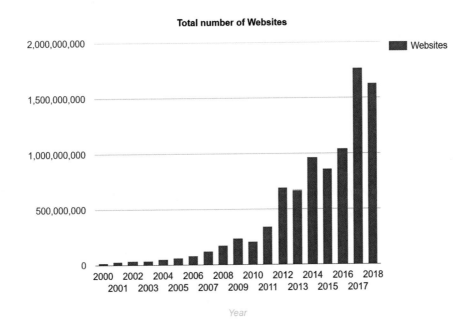

Fig. 7.1 Worldwide website growth in the past 20 years (*Data Source* internetlive stats.com)

search engine results pages (SERPs). The information may be the combination of hyperlinks (URLs) to web pages. which contain images, videos, and even animations of any multimedia information. Some search engines can even mine data available in databases or open directories and repository. Beside textual search, search engines such as Google even provide image search—web search by images or photos. It is different from web directories which are maintained by human editors only, these search engines also maintain real-time information by running an algorithm on a web crawler to be studied in the following section.

7.2   A Brief History of Search Engine

1993:   The first widely acclaimed search engine, *Web Wanderer* appeared. Created to measure the growth of the web, it performed its job through 1997. Statistics compiled by this search engine are still available on the web today.

1994:   *WebCrawler* came into the scene. The original *WebCrawler database* contained 6,000 websites only. *AOL* bought *WebCrawler* in 1995 but sold it to *Excite* two years later. *Infospace,* its current owner, bought *WebCrawler* when *Excite* declared bankruptcy.

1994:   Another powerhouse, the *Lycos search engine,* was launched with 54,000 indexed documents. The *Lycos search engine* is still a player today but changed hands several times. Currently is a subsidiary of Indian-based company *Ybrant Digital.*

1995:   *AltaVista* was the first search engine to include multilingual search capabilities. *AltaVista* became *Yahoo's* property eventually (discussed later). *AltaVista* was the search king until *Google.* Their services shut down in 2013 and domain redirected to *Yahoo's* own search site.

1998:   *Alphabet Inc.* co-founders *Larry Page* and *Sergey Brin* introduced *Google* to the world which quickly jumped to the top of search engine rankings. The name came from the word *googol,* which is the name for number 1 followed by 100 zeroes.

2006:   *Google* became a part of our culture that several dictionaries *Merriam-Webster* and *Oxford English Dictionary* included *google* as a verb.

2009:   *Microsoft* launched *Bing* to introduce the use of suggested searches along with search results.

2014:   More searches were performed using mobile devices than using desktop browsers for the first time in history.

2015:   Over half of *Google* searches were performed using mobile devices. *Google* announced that they would commence using mobile-friendly factors in its mobile search results, meaning they list sites that were not mobile-friendly lower in mobile searches. Figure 7.2 shows worldwide search engines ranking as of today.

Google is the world's most used search engine as of 2020 with a market share over 90%. As one can see, it is still the most used search engine on the Web since the very beginning. *Google* now enjoys greater than 50% of total search engine traffic. This means that a top *Google* ranking will yield more traffic to the site than a top ranking with any other search engine. *Google's* popularity is due largely to speed and search results quality. Both are possible because of a worldwide network of more than 1 million servers which house *Google's index*. The sheer number of servers and speed at which they communicate with each other is unparalleled in the search industry. With the advancements in AI, deep networks, cloud computing, and big data, today's competition between search engine corporations become even more challenging and severe than before. (Haider and Sundin 2019; Croft et al. 2009).

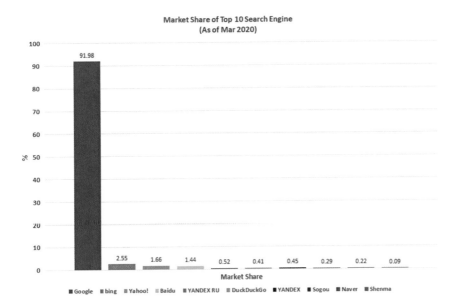

Fig. 7.2 Market share of top 10 search engine as of March 2020 (*Data source* statco unter.com)

7.3   Main Components of Search Engine

Although different search engines have their own system architecture, basically, a typical search engine consists of five main components: (1) *Crawler Module*; (2) *Page Repository Module*; (3) *Indexing Module*; (4) *Query Module*; and (5) *Ranking Module*. Figures 7.3 shows the main components of a classical search engine.

Crawler Module

A *web crawler*, also known as a *web spider, spider bot, web bot*, or simply a *crawler* are computer programs that scan the Web, *read* everything they find (Gunjan and Subramanyam 2014).

A typical web crawler scans web page to see what words they contain and where those words are used. The *crawler* then turns its findings into a giant index and stored into the index database for further processing. Compared with traditional document collections which reside in physical warehouses such as college library, the information available on the Web is distributed over the Internet.

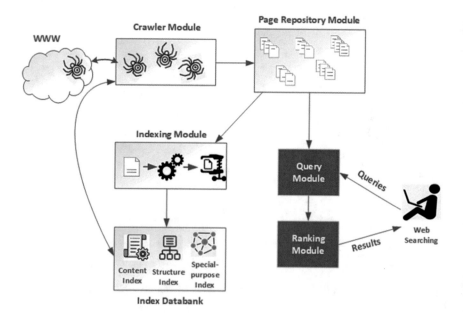

Fig. 7.3 Main components of a classical search engine

In fact, this huge *repository* is growing rapidly without any geographical constraints. Hence, a component used *crawler* is employed by a search engine which visits web pages, collect, and categorize them. The truth is, one of the major aspects for different search engines performance is the algorithm being used in the *crawler program*. In other words, *crawler module* is the most important module for the search engine nowadays.

Page Repository Module

The downloaded web pages are stored temporarily in a search engine local storage called *page repository*. The new pages remain in the repository until they are sent to the indexing module where their vital information is used to create a compressed version of the page. The retrieved web pages contain various types of multimedia information including images, photos, videos, and even embedded programs instead of textual information most of the time. For most search engines, these additional resources are either discarded or temporarily stored in a separated repository store in order not to affect the system performance of the search engine.

Indexing Module

The main function of the *indexing module* is to process each newly uncompressed web page from *page repository*, extract suitable descriptors, and then generate a compressed description of the web page. The compressed web page is stored in a databank, which can be accessed through a browser if necessary. The index databank is used to store all the compressed information for each web page being crawled.

There are basically 3 index types: *content index, structure index,* and *special-purpose index*. The *content index* is the most important index that contains all keywords, title, and anchor texts being crawled in each web page and they are stored in a compressed form using an inverted file structure. The *structure index* is used to store the link information of the web page also in compressed form. The *special-purpose index* stores all nontextual indexes such as photo, images, multimedia, and even pdf file indexes, which will be used for specific query tasks, e.g. image search in *Google*. *Web crawlers* crawl the Web constantly, bring back new and updated pages for indexing and storage. These 3 modules and their corresponding data files are operated independently to the queries sent by web users.

Query Module

The main function of the *query module* is some sort of NLP module to process the user's search query from natural language query type into the format that can be comprehensible by search engine and consult various indices in order to answer the query. For instance, the query module consults content index and its inverted file to find which pages contain the query terms, i.e. the relevant pages which are then passed to the ranking module.

Ranking Module

The main function of the *ranking module* is to rank web pages according to the *matching degree* of the user's query. The *ranking criteria* in traditional search engines include: the *term frequency, popularity score, content score,* etc. After ranking process is finished, the ranking module will generate an ordered web pages list according to matching degree. The ranking module is in fact the most important process to evaluate whether a search engine is *good* or *bad*.

A search engine by common sense is *good* not only because it can search and results at speed. More importantly, it should show how close result web pages are related to the user's query.

When a user tries to search via search engine every time, there will be over thousands or sometimes up to millions of query results in reality. So, how to rank query results is a critical issue to determine whether a search engine is effective or not. The page ranking is computed using rules by combining two scores; the content score and the popularity score. For example, many web search engines give pages, using query word in the title, as a higher content score as compared to pages containing the query word in the page body. The popularity score is determined from the analysis of Web's hyperlink structure. The content score is combined with a popularity score to determine an overall score for each relevant page. The set of relevant pages resulting from the query module is then presented to the user in order of overall scores.

7.4   Google Crawler

Google search engine uses multiple machines for web crawling to speed up its search engine efficiency. *Google search engine* crawler process consists of five components (Rastogi et al. 2013; Lee et al. 2019):

(1) URL server in *Google search engine* reads URLs out of a file and forward them to multiple crawler processes.

(2) Each crawler process operates on a different crawling machine to speed up the crawling process. It uses asynchronous I/O to fetch data from up to 300 web servers in parallel in a typical multithreading crawling process normally as shown in Fig. 7.4.

(3) The *crawlers* transmit downloaded pages to a single *store server* process, which compress the pages and store them to disk.

(4) The *indexer* process reads web pages from index storage, then extracts links from pages and saves them in a separate index storage.

(5) A *URL Resolver* process reads the link file, analyzes URLs contained therein, and saves the absolute URLs to the disk file read by the URL server.

Fig. 7.4 Google Crawler with multiprocessing operations

Besides, Google search engine uses multiple crawlers (so-called *bots*) to find different types of information. They are.

- *Googlebot*—to index content for showing in Google web search results. This is also the same crawler used for mobile phones.
- *Googlebot-Image*—to index images for showing in Google image search results.
- *Googlebot-News*—to collect news feed for showing in Google News search results.
- *Googlebot-Video*—to crawl videos on the Web for showing in video search results.
- *Googlebot-Mobile*—for Google mobile search on feature phones.
- *Google-Mediapartners*—for indexing web page content to display relevant Google AdSense ads.

7.5   Search Engine Architecture

7.5.1   Google Search Engine

The word *Google* comes from the word *googol,* which stands for *10,100.* Google search engine, www.google.com, claims to be the best search engine today. The entire *Google engine* is written in C/C++ and operated under *Solaris/Linux platforms* to ensure high performance (Rastogi et al. 2013; Lee et al. 2019). Figure 7.5 shows its system architecture.

The entire Google search engine operation includes.

- Google URL sends URLs lists to be fetched by Google crawlers (with multiple crawlers working simultaneously).
- Google crawlers download web pages from the URL list and send the download pages to the Storage Server.
- The *Storage Server* compresses web pages and stores them into the repository.
- Every web page being fetched is associated with a unique ID called *docID* with an index.
- The search engine reads the repository, uncompressed documents and performs the parsing operation.
- Each page is converted into a set of word occurrences, so-called *hits*.

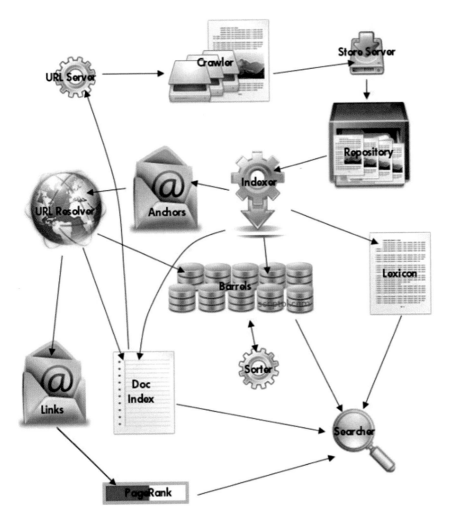

Fig. 7.5 System architecture of Google search engine

- The *indexing engine* distributes these hits into a set of *barrels* and creates a sorted forward index partially. It parses out all the links in every web page and stores important information about them in an anchors file.
- The URL resolver reads the anchors file and converts relative URLs into absolute URLs and in turn into docID.
- The Google ranking database is used to compute Page Ranks for all documents.
- The Google Sorter takes the barrels and resorts them by wordID instead of docID to generate the inverted index. Also, the *sorter* produces a wordIDs list and offsets into the inverted index.

- The *Google Lexicon* takes this list together with lexicon produced by the indexer and generates a new lexicon to be used by the searcher.
- The *Google Searcher* uses the lexicon built by Google Lexicon together with the inverted index and the Page Ranks to generate answer queries and send back to the user's browser.

7.5.2   Google Image Search

An image search engine is an images database that can be searched through keywords, so that users can find valuable images. *Google's image search engine* (Andrew 2018) was created on July 12, 2001 and is one of the most comprehensive and famous image search engines today. Now we can explore the Web in an entirely new way by beginning our *Google search* with an image. There are a few ways to search by image. We can visit *images.google.com* or any images result page and click the camera icon in the search box. Enter an image URL for an image hosted on the Web or upload an image from our computer. Search by image works best when the content is likely to show up in other places on the Web. For this reason, it is likely getting more relevant results for famous landmarks or paintings than personal images. Figure 7.6 shows an example of *Google image search* for "*orchids*".

Here are the top 5 image search engines in the market:

(1) *Google image search.*
(2) *Yahoo image search*—Yahoo image search comes in second place as one of the best image search engines because a user can find great images for every niche. When a user types niche into the search bar with a variety of images, they can also find professional lifestyle images, images with white backgrounds, graphics, and more from the store.
(3) *Bing image search*—Bing image search is alike Google Images and Yahoo Images so is another great picture search engine to source images. Type user niche or what s/he needs an image into the Bing Image search bar.
(4) *PicSearch*—PicSearch has a collection of 3 billion images and is no wonder one of the best image search engines. The image search engine sources pictures from a variety of websites. User can select from lifestyle pictures to stock photos. But the user still needs to obtain permission to use images on the platform as per the disclaimer at the footer of the website.
(5) *Yandex*—Yandex is another free website with a large images database to search. It is the most popular search engine in Russia but small compared to its competitors. Yandex only shows where images being used with a list

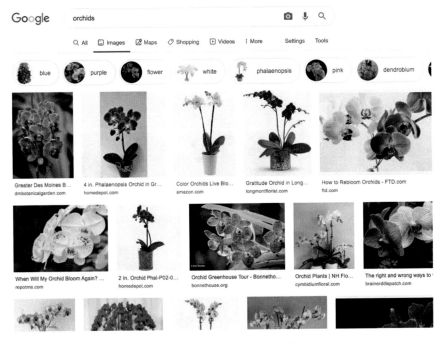

Fig. 7.6 Example of *Google image search* for "orchids"

of different sizes. If the database has nothing to satisfy the exact search it will show similar images to the user.

7.5.3   Harvest Search Engine

There are several variants of *crawler-indexer*-type search engine architecture in addition to *Google Search Engine*. One important variant is the so-called *Harvest architecture* (Büttcher et al. 2010; Rastogi et al. 2013). *Harvest* is a system to collect information and make them searchable using a web interface by using distributed architecture to gather data and distribute data. It was used by US Government departments and national institutions as their official search engine originally.

Netscape's Catalog Server is a commercial version of *Harvest and Network Appliances'* cache. Netscape is also in fact the first and one of the most popular commercial browsers used in the 1990s. Figure 7.7 shows the system architecture of the Harvest search engine.

Harvest search engine consists of two main components: *Gatherers and Brokers*. It is different from traditional search engine using the so-called *crawler-indexer* model. The role of *Gatherers* is to collect and extract indexing

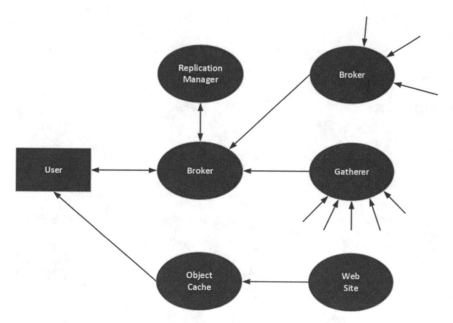

Fig. 7.7 Harvest search engine system architecture

information from one or more *web servers* in which the gathering times are specified by the *Harvest Search Engine*. The role of *Brokers* is to provide indexing mechanism and query interface to data the system gathered, the brokers receive information from the *Gatherers* or other *Brokers* in general to update their indices. Besides, *Brokers* will also filter information and send to other *Brokers* to save searching time. By regulating *Gatherers* and *Brokers* configuration, *Harvest Search Engine* workload can be balanced. *Harvest System* builds topic-specific brokers and focuses index contents by avoiding many of generic indices vocabulary and scaling problems in general. For example, the system provides a replication manager and object cache to reduce network and server load.

7.5.4 AltaVista Search Engine

AltaVista is the first search engine to index every word on a page and provides a retrieval system to extract relevant information (Seltzer et al. 1997; Rastogi et al. 2013). It was developed by *Digital's Research Labs* in Palo Alto in 1995. AltaVista search engine was available in more than 25 languages in a variety of versions. AltaVista was acquired by *Overture* in 2003 and by *Yahoo* later in the year. AltaVista search was redirected to Yahoo in 2011.

The crawler's duty in *AltaVista Search Engine* is to run on a local machine and sends requests to remote *web servers*. The index is used in a centralized fashion to answer queries from users. Figure 7.8 shows the system architecture of the *AltaVista Search Engine*.

AltaVista search engine system architecture consists of two components. The first component consists of the user interface and query engine. The second component contains *crawler and indexer*.

AltaVista was running on 20 processors in 1998. All processors have 130 GB RAM and over 500 GB hard disk space. Only the query engine occupies more than 75% of these resources. There are two problems with this architecture. The first problem is data gathering in the dynamic web environment, which uses saturated communication links and high load at web servers. The second problem is data volume. The crawler-indexer architecture does not cope with web growth for the future.

By the time AltaVista was considered a serious search tool, its focus was changed to a portal, and users were not fond of it. While acquisitions and financial struggles had been playing out to *AltaVista*, *Google* started to gain ground. Additionally, *Google* foresaw a problem with spam and low-quality search results. Many experts believe that Yahoo was rather hasty in its decision to close AltaVista without fanfare. But by the time *AltaVista* was closed in

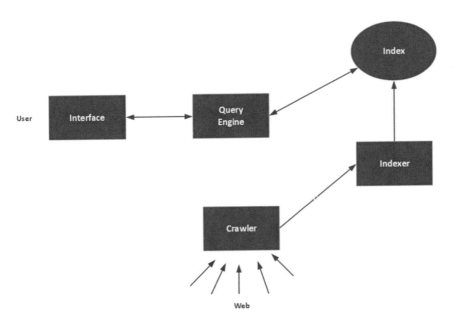

Fig. 7.8 *AltaVista search engine* system architecture

2013, the internet community had moved on; users that mourned its demise almost certainly did not use it any longer.

7.6   Major Problems of Tradition Search Engines

7.6.1   Search Performance Perspective

As said, traditional search engines led by Google did provide us with great assistance to search for information and websites for the past 20+ years. However, owing to the ever-increasing of (1) over thousands to millions of new websites on a daily basis; (2) users increase demand for search engine expectation, existing search engines face many new challenges which include.

(1) *Capability*—as mentioned at the beginning of the chapter, traditional search engines such as *Google, Bing, or Baidu* are all based on *keywords* concept. In terms of linguistics, such keyword search is only restricted to NLP basic *syntax level*. For example, if a user types in a search query: *I want to see a jaguar*. Does it mean a real animal jaguar or the car brand *Jaguar*? It will probably show car brand Jaguar either because it is the ads related to keyword *Jaguar* or the most *popular vote* for using keyword *Jaguar* to search, without any *semantic-level* or even *pragmatic-level* understanding what the user's means in their search query or previous browsing history. Figure 7.9 shows the search results for using query: *I want to see a jaguar* in *Bing Search Engine*.

(2) *Efficiency*—Traditional search engines are relatively slow to index dynamic real-time data such as tweets or social media comments. This problem is twofold: (1) *The purpose*—traditional keyword-based search engines such as Google are tailored for the fetching information from websites and their underlying web pages originally, which are somehow more static than the tweets or social media comments in a real-time update; (2) *The volume*—new tweets or social media comments are another dimensional volume compared to websites, especially if the search engine is not *intelligent enough* to decide which real-time tweets need to be searched or not, or simply do it in a random manner.

(3) *Coverage*—Search engines index web pages but users often want the underlying data, e.g. stock prices, flight times, weather reports, etc. In other words, users today are using search engine not only to search for some websites or information, but also rather would expect the search

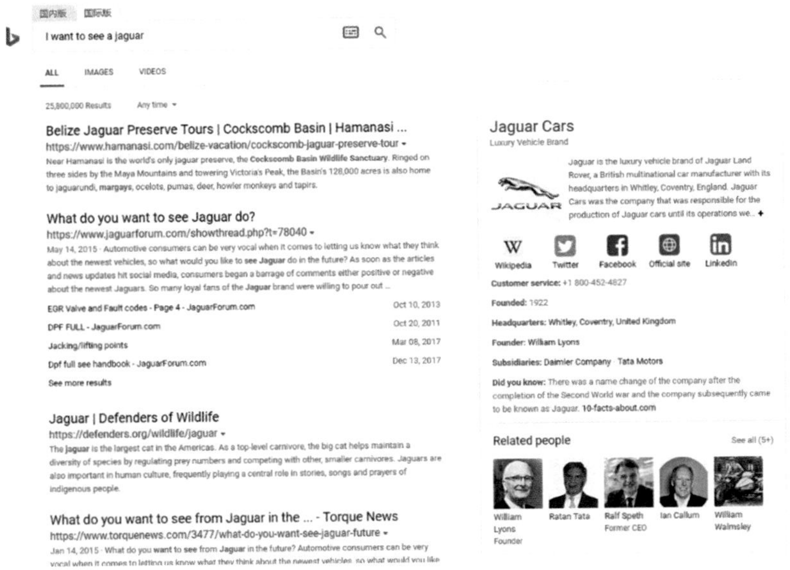

Fig. 7.9 Ambiguous of traditional search results

engine to *understand* what they really meant to provide the answer or information they want to know.

7.6.2 User Perspective

There are other problems for users apart from traditional search engines system performance concerns (Büttcher et al. 2010; Rastogi et al. 2013). They include.

(1) Most users do not understand exactly how to provide a search query. This problem in fact has two aspects: (1) All popular search engine are all keyword based. Users need to provide *keywords* that related the information they want to search to provide better search results; (2) Not everyone knows the keyword related to the information they want to search. They simply type the complete query string in natural language sentence.

(2) Users have difficulties to understand *Boolean logic* resulting in unable to perform advanced searching. Since existing search engines all followed logical inference for searching. Users will not able to perform if they do not understand *logical inference* or *logical operations*.

(3) Users do not care about advertisements resulting in a lack of funding. This issue is a chicken-and-egg problem. The main reason for popular

search engines free-of-charge today is that they believe users will give up using if they are not free. The reason why all commercial search engines are full of ads especially at the top query result list will affect the accuracy and creditability significantly for a search engine to provide the *best match* results for users, instead of the *best related commercial ads for their queries*.

(4) Around 85% users only look at the first-page result, so relevant answers might be skipped and ignored. Search engines today have one intrinsic problem, as all the search results are searched automatically and ranked according to similar query keywords. The first-page query results are similar in contents usually and the overall query pages are numerous usually. Users simply browse the first (or second) page and give up. But maybe the *true results* appeared in the later pages because they are closely related in *context*, but not in *keyword hits* or *term frequency*. Figure 7.10 shows the *best SUV* query results in Bing, as one may see, all top query results are ads. Are they really the best SUV, or just the ones that pay most?

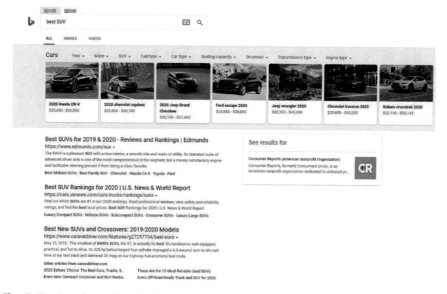

Fig. 7.10   Search results of *"best SUV"*

7.7   Knowledge and Ontology

7.7.1   From Knowledge to Computational Knowledge

What is Knowledge?

Since *Ancient Greece*, many philosophers have been discussing *knowledge*. It is not easy to highly abstract knowledge definition (Abbas 2010; Effingham 2013). *Knowledge* can be defined as meaningful resources that make us understand the world. *Knowledge theory* defines the content about the world, how to encode it, and how we reason about the world. Unless we define it for the computer processing purpose rather than human understanding, similar definitions can be applied to computers and information systems. This is called *computational knowledge* (Lim et al. 2011).

Computational Knowledge

In many *Knowledge Management Theories* (Denning and Tedre 2019), *Computational Knowledge* (CK) in computer science can be regarded as hierarchical: *data→ information→ knowledge*. Figure 7.11 shows the famous *Knowledge Pyramid*. In this model, *data* refers to original data consisting of strings of digits, numbers, or symbols that are meaningful only to the program. *Information* refers to combining data to produce a certain degree of basic meaning. *Knowledge* refers to gathering information to provide practical

Fig. 7.11 The knowledge pyramid

knowledge. *Wisdom* refers to the highest level of knowledge corresponding to the generated insights and wisdom, which is our ultimate knowledge goal.

Creating *computational knowledge* is AI study. Advanced information systems, such as information retrieval systems, forecasting systems, resource management systems, online shopping systems, personalized systems, etc. always require *computational knowledge* to perform tasks with higher intelligence. Traditional information systems lack intelligence because they process data and information without analyzing the knowledge behind them. We need to discover and represent knowledge from raw data to computable forms and express it for processing in order for computers to understand and process knowledge. Intelligent information systems with the ability to process knowledge are called *knowledge-based systems (KBS)*.

7.7.2   Knowledge Engineering and Representation

Knowledge engineering (Brachman and Levesque 2004) has evolved rapidly over the past decade as the demand for knowledge-based systems has grown. Knowledge engineering is a process of extracting useful knowledge from computer data. It requires the process of analyzing and discovering data patterns and converting them into a format understandable by humans or computers or both. Over the years, knowledge engineering research has been devoted to theories, methods, and software tools development to help individuals acquire computer knowledge. They use scientific and mathematical methods to discover knowledge. These methods can be simply defined as *input–process–output* systems: *Input*—computer data sets, such as text and database records; *Process*—a method of converting input data into knowledge; *Output*—in a specific form of knowledge representation (such as ontology). Knowledge representation language is used to express knowledge in information systems. It can be classified according to the original type used by the user (Guarino 1995) as shown in Table 7.1.

They are also divided into five different levels:

1. *Logical level*—contains basic primitives including predicates and functions. It also allowed formal primitives interpretation.
2. *Epistemological level*—is a knowledge structure used to fill the gaps between logical levels (conceptual levels are specific conceptual meanings models), which are general and abstract primitive elements.
3. *Ontology level*—is an ontology commitment that includes ontology relationships related to clearly specify language primitives.

Table 7.1 Knowledge representation formalisms (Guarion 1995; Lim et al. 2011)

	Level Type	Primitives	Interpretation	Main features
1	Logical	Predicates, functions	Arbitrary	Formalization
2	Epistemmological	Structuring relations	Arbitrary	Structure
3	Ontological	Ontological relations	Constrained	Meaning
4	Conceptual	Conceptual relations	Subjective	Conceptualization
5	Linguistic	Linguistic terms	Subjective	Language dependency

4. *Conceptual level*—contains primitives with a clear cognitive interpretation, corresponding to concepts meaning not related to language.
5. *Linguistic level*—contains and process primitive language terms related to language-related nouns and verbs.

7.7.3 What is Ontology?

Ontology (Effingham 2013) originated from philosophy and has grown into a popular research in computer science and information systems. From philosophy perspective, ontology is a philosophical study of existence. More broadly, it studies concepts that are directly related to *existence*, especially to existence, reality, and the basic categories of existence and its relationship. Traditionally, ontology refers to a part of the main branch of metaphysics philosophy. Ontology usually deals with the existence and arguably of entities existence and how to group these entities, to associate and subdivide within the hierarchy according to similarities and differences between them.

For example, two famous philosophers in human history, Aristotle and Kant believed that ontology is a *study of existence*. It refers to a *category system* used to describe *existence* or *classification* of real worlds. Although ontology concept has not been developed in Aristotle's philosophy in Ancient Greece, his classification system is still related to the definition of the ontological classification system today. Table 7.2 shows the ten categories of Aristotle's expression of things or existence (Aristotle 2014).

Immanuel Kant (1724–1804) proposed a new classification system (Willaschek 2018) in 1781. The system is divided into four categories, and each main category is divided into three subcategories as shown in Table 7.3. This classification system and category are also related to ontology development today, especially the ontology is highly dependent on describing the

Table 7.2 Aristotle's ten categories (Aristotle 2014; Lim et al. 2011)

	Categories	Descriptions
1	Substance	What or being
2	Quantity	How much
3	Quality	What kind
4	Relation	With respect to
5	Place	Where
6	Time	Whe
7	Position	To lie
8	State	To have
9	Action	To do
10	Affection	To undergo

Table 7.3 Kant's categories (Kaye 2019; Lim et al. 2011)

	Categories	Sub-categories	Descriptions
1	Quantity	Unity	Universal
		Plurality	Particular
		Totality	Singular
2	Quality	Reality	Affirmative
		Negation	Negative
		Limitation	Infinite
3	Relation	Inherence and subsistence	Categorical
		Causality and dependence	Hypothetical
		Community	Disjunctive
4	Modality	Possibility or impossibility	Problematical
		Existence or nan-existence	Assertoric
		Necessary or contingence	Apoditic

relationship between an entity or existence. The subcategories of this reference can be regarded as different types of objects, attributes, and ontology relationships.

Ontology is also the basic form of *knowledge representation* about the real world. From the computer science perspective, *ontology* defines a set of representational primitives used to model *knowledge* or *discourse domains*. The ontological representation primitives contain classes, attributes, and relationships between classes. They are used to model specific domain knowledge. Ontology is helpful for knowledge-based systems (KBS) development, thereby achieving knowledge sharing and reuse. It enables intelligent communication between computers, such as communication language used in *intelligent agents (IA)*. Standardized specifications allow knowledge engineers to develop their own *ontology* by reusing and sharing with each other.

7.7.4 Computational Ontology

Computational ontology is usually expressed in different abstract forms: *top-level ontology, lexical ontology, and domain ontology* (Gruber 1995; Lim et al. 2011; Mascardi et al. 2010).

Top-Level Ontologies

The top-level ontology (also called the *upper-level ontology*) is limited to *universal, generic, abstract, and philosophy concepts.* They are general enough to handle high-level abstractions and a wide domains range. The concepts defined in the top-level ontology are not specific to a specific domain, e.g. medicine, science, finance but only provide a structure and a set of general concepts from which the domain ontology can be constructed. This top-level ontology promotes data interoperability, semantic information retrieval, automatic reasoning, and natural language processing. The *Standard Upper Ontology Working Group* (IEEE 2020) has developed upper ontology standards for computer applications for data interoperability, information search and retrieval, and natural language processing (NLP).

Upper-layer ontologies examples include SUMO (recommended upper-level merged ontology) (SUMO ontology 2020), CYC Ontology (OpenCyc 2020), and SUO 4D Ontology (SUMO ontology 2020). SUMO has been proposed as an entry document for the SUO working group. It creates a hierarchy of top-level things as entities and contains physics and abstraction. SUMO divides ontology definition into three levels: upper-level ontology (SUMO itself), middle-level ontology (MILO), and lower-level domain ontology. The intermediate ontology acts as a bridge between the upper abstraction of the domain ontology and the rich details of the lower layer.

OpenCyc (OpenCyc 2020; Lim et al. 2011) is one of the most complete common-sense knowledge bases that use upper-level ontologies. Today, it has been regarded as a standard upper ontology by IEEE. It is some formal common sense. It models common sense libraries and aims to solve common sense problems. The entire *Cyc ontology* contains hundreds of terms, and these terms have a mutual relationship to simulate human consensus reality. It contains a knowledge server that provides services for its *Cyc knowledge base*, an inference engine, and defines a *CyCL representation language* for knowledge representation. It is an upper ontology that can be used to define some lower-level ontology knowledge such as *domain-specific knowledge*, domain-specific facts and data (OpenCyc 2020). Figure 7.12 shows *OpenCyc's top-level ontology* hierarchy.

Fig. 7.12 OpenCyc top-level ontology structure (Opencyc 2020; Lim et al. 2011)

Lexical Ontologies

Lexical ontology is an ontology that describes linguistic knowledge. It attempts to model the words meaning through ontology structure. Such ontology examples include *WordNet* (Cognitive Science Laboratory 2009; Miller 1998) and *HowNet* (Dong and Dong 2006).

WordNet was originally designed as a *lexical database* of English words (Miller 1998). It can be used as a lexical ontology to represent knowledge used for computer text analysis and AI application development, especially for many applications related to natural language. *WordNet* defines *synsets* to group English nouns, verbs, adjectives, and adverbs into synonym sets, and uses different grammatical rules to distinguish them (noun verbs, adjectives, and adverbs). It is helpful to model words concept and their semantic relationship. It has been used in various natural language text analysis such as word sense calculation and ambiguity elimination. WordNet research has been extended to *ImageNet*. It uses meaningful concepts in WordNet to connect to image data. This is a practical example of using WordNet as knowledge to build an intelligent information system (concept-based image database). Figure 7.13 shows a word ontology example in WordNet.

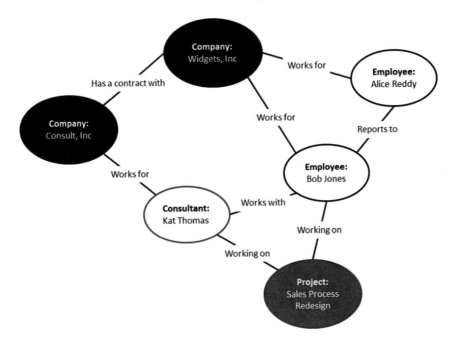

Fig. 7.13 Example of *Lexical Ontology* in WordNet

HowNet (Dong and Dong 2006; HowNet 2020) is a Chinese word *lexical database*. It covers more than 65,000 Chinese concepts, equivalent to about 75,000 English concepts. It is common sense knowledge used to model the relationship between concepts and attributes between Chinese lexicon concepts and their equivalents in English (Dong and Dong 2006; HowNet 2020). HowNet uses its structural knowledge based on Chinese words for Chinese processing. Alike to WordNet's *synset*, HowNet defines its specific Sememe-network to model the inter-concept relationship between Chinese lexicon concepts. HowNet is a fully computable electronic database. HowNet's knowledge consists of graphics. Figure 7.14 shows a graph-based example used to describe different concepts, attributes, and their interrelationships. HowNet is also defined a taxonomy as a high-level ontology for Chinese lexicon conceptual model category.

Although a lot of work has been done on these lexical ontologies and they have also conceptualized lexical knowledge very effectively. The main problem with these ontologies is that they are created manually throughout the process. The disadvantage of manual processing is that it is difficult to maintain, i.e. add new knowledge, modify, and update existing knowledge. The concept and words usage are constantly changing, so defined

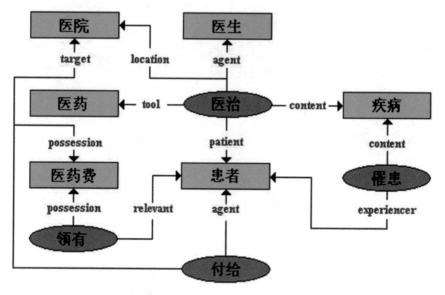

Fig. 7.14 Example of lexical ontology in HowNet (Lim et al. 2011)

words knowledge is not effective permanently. Therefore, continuous manual update work and reconstruction are required making the process ineffective.

Domain Ontologies

Domain ontology (Simiraglia 2015) is related to specific domains that can be extended from *upper ontology*. It should be defined for a specific domain, because even a huge ontology like Cyc contains more than modeling general and high-level 10,000 concepts, but its depth is not enough to express domain-specific concepts and low-level concepts. (For example, medicine, science, finance, etc.). A domain ontology is formed based on concepts in specific domains of interest in order to model domain knowledge and make its information expressible and understandable by machines. *Domain ontologies* are preferably constructed based on available upper-level ontologies such as Cyc to facilitate mapping and integration between different domain ontologies created by different experts or researchers to enhance sharing and usability.

Unlike upper-level ontologies that are frequently used to infer common-sense knowledge, domain ontologies are mainly used to reason about specific knowledge domains. Domain ontology is a boarder and more general in defining knowledge. In other words, the domain ontology is not abstract but

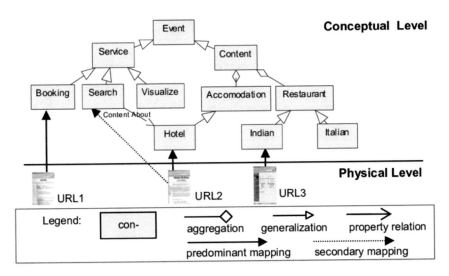

Fig. 7.15 Example of domain ontology (Simiraglia 2015)

more specific. Therefore, building smart applications is more useful because computer applications are usually developed for specific target domains. Figure 7.15 shows the ontology tree specified for entertainment (part of the entertainment domain). Most application ontologies rely on domains, but they share with each other across different domains. Ontology engineering usually aims to define and create domain ontology, not top-level and lexical ontology. In fact, ontology engineering's main research and development is also focused on domain ontology.

7.7.5 Ontology Engineering Tools

Ontology engineering tool or *ontology editor* (Simiraglia 2015; Lim et al. 2011) is an application that can help ontology engineers and domain experts create and maintain ontology. These applications operate ontologies in different ontology languages, providing ontology management functions on the screen, i.e. create, edit, verify, import, and export. Examples of such applications include Protégé, Onto-Builder, OntoEdit, Construct, etc.

  Protégé is a free open-source ontology editor for creating domain models with ontology and knowledge-based information systems. It was developed by Stanford University. The Protégé editor supports various ontology languages, such as RDF, RDF(S), DAML + OIL, XML, OWL, Clips, UML, etc. Protégé supports two main ontology modeling methods include:

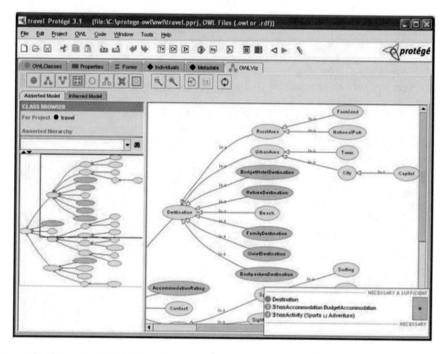

Fig. 7.16 The Protégé-OWL editor for OWL ontology modeling (Lim et al. 2011)

Frame-based ontology modeling conforms to the *open knowledge base connection.* The model consists of a set of classes (ontology taxonomy) representing domain concepts in ontology hierarchy, a set of association slots (ontological relations) in classes representing attributes and relationships, and a set of instances (instantiation) of these classes.

An ontology model based on OWL (OWL 2020) for semantic web (W3C Semantic Web 2020). OWL is a language for describing classes, attributes, and their instances. It officially specifies how to derive the logical results of ontology and aims to develop semantic web applications. Figure 7.16 shows a Protégé-OWL editor screenshot used for OWL ontology modeling.

7.8   Semantic Web

7.8.1   What is Semantic Web?

The current web system is mainly based on HTML. HTML was designed originally for information exchange and display only. Therefore, they are not designed to understand web content for themselves. Enriching web content with semantic data aims to solve this problem. The *semantic web* (Allemang and Hendler 2011) is a type of technology that adds more structured markup data to the semi-structured information of HTML pages. This semantically marked data benefits from the machine's legibility. Therefore, it can enhance proxy applications to handle web content. There is also a close relationship between *ontology* and the *semantic web* because ontology is a key element in constructing semantic web content.

The *semantic web* is not only used to provide web data for human use, but also to create data that machines can *understand and process*. The main vision of the *semantic web* is to create data that machine can process and define how the machine operates on the data and make the web system more intelligent. We need a semantic network because today's network information is overloaded. Because web data amount is too large to be used by human, we need machines to help us process a large information amount before delivering it to us. Information processing such as information filtering, searching, and recommendation requires high-level machine intelligence, and *semantic web* technology enables us to develop such intelligent systems effectively. The adoption of semantic web technology can benefit many organizations in current business processes and improve their efficiency. Daconta et al. (2003) described the idea that semantic web can take advantage of the organization's best interests and modified it as shown in Fig. 7.17.

Semantic web has knowledge as its core component. It is located in data that machine can process. These data enable the machine to analyze and return some useful results to the user. The results of these analyses are helpful for decision-making and marketing purposes. The machine can even provide users with certain expert advice or suggestions and provide more valuable

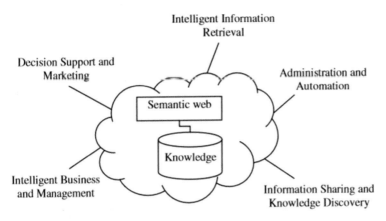

Fig. 7.17 Semantic web and KnowledgeBase (Sheth 2013)

knowledge to help decision-making. Traditional business applications such as e-commerce and customer relationship management only provide static data, such as product information, transaction records, customer information, etc. Semantic web has reasoning capabilities that can be used in e-commerce docking. It helps to connect potential customers with business partners or sales departments. This intelligent business and management function create more profit opportunities for the organization.

7.8.2   Semantic Web Framework

This section introduces the semantic web defined by W3C (W3C Semantic Web 2020), which involves basic concepts and technologies that support semantic web development. Figure 7.18 visualizes semantic web stack through W3C. It is divided into different layers and semantic web can be developed.

It uses a self-describing markup language XML starting from the bottom which can exchange data on the entire Web, but it does not mean any meaning and knowledge embedded in the data. Therefore, RDF (Resource Description Framework) and RDF schema are defined and built on top of XML. It can be used to model abstract data meaning and data semantics representation. The data semantics in RDF (based on XML) can therefore easy to handle and understand by software agent. Finally, model the

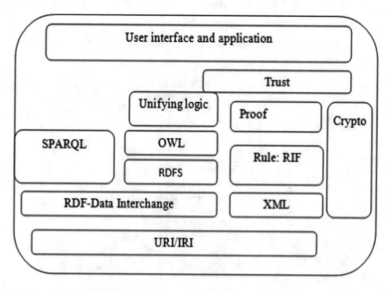

Fig. 7.18   Semantic web framework (Lime et al. 2011; W3C 2020)

ontology knowledge in OWL. OWL defines more detailed information about attributes, classes, relationships between classes, cardinality, equality, etc. SPARQL defines a query language for semantic web data. This includes the lower layer (data layer) in semantic web stack. The upper layers of semantic web architecture include proof and trust. It describes distributed data accessibility and credibility. Web applications can reason about derived results confidence based on these layers.

7.8.3 Semantic Modeling

Semantic modeling in information technology refers to mapping or formalizing human knowledge into a certain language syntax (Allemang and Hendler 2011). Human knowledge is expressed in unstructured natural language usually which is difficult for computer processing. Therefore, we need some structured language grammar to model potential *semantics* behind the natural language. The semantic modeling main idea is to associate the term in the sentence with a concept in the real world that the term refers to. Various techniques have been developed to handle semantic modeling tasks. We have simplified these semantic modeling techniques from weak semantics to strong semantics based on our ability to express knowledge (Lim et al. 2011). Figure 7.19 shows the 4-level structure of semantic modeling.

Taxonomy

Taxonomy (Allemang and Hendler 2011) describes knowledge in terms of hierarchical structure or *parent–child relationship semantics*. Taxonomy is a type of classification system in the form of relationship between categories

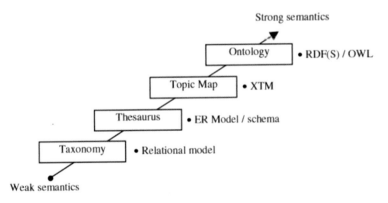

Fig. 7.19 Semantic modeling structure (Lim et al. 2011)

and subcategories. A typical taxonomy is the animal classification in biology. For example, animals are classified as cartilaginous animals, arthropods, mollusks, etc., while *carbonates* are further classified as Aves, reptiles, amphibians, mammals, etc. and mammals include human, cats, dogs, etc. It is useful when describing creatures in the real world and has a long history in biology. *Taxonomy* can also be found anywhere in the information technology environment such as the folder structure in computer drives and websites site map. For example, financial website content can be classified as investment, news and experts, personal finance, etc. and investment can be further classified as today's market, market events, etc. Finally, today's market contains hyperlinks to the market, overview, market updates, etc.

Thesaurus

Thesaurus can be defined as: A *controlled vocabulary arranged and organized in a known order so that the equivalence, isomorphism, hierarchy, and association between terms are clearly displayed and identified through standardized relationship indicators* (Lim et al. 2011). Therefore, it can describe knowledge better than taxonomy. The relationship between the terms in controlled vocabulary is used to associate the meaning of the term with the meaning of other terms. WordNet (2020) is an example of English thesaurus and Chinese *HowNet* (HowNet 2020). Table 7.4 shows different types of semantic relations and their examples.

Table 7.4 Examples of semantic relations in thesaurus (Lim et al. 2011)

Relationship type	Example
Equivalence	
Synonymy	"HK"/"Hong Kong"
Homographic	
Homonym	"Mouse" (animal)/"Mouse" (input device)
Hierarchical	
Hypernym	"Mouse"/"Mammal" (child-of)
Hyponym	"Mammal"/"Mouse" (parent-of)
Meronym	"Window"/"House" (part-of)
Holonym	"House"/"Window" (has-part)
Associative	
Cause-effect	"Accident"/Injury"
Attribute-host	"Color"/"Cloth"
Material product	"Grapes"/"Wine"
Location-event	"Hospital"/"Medical treatment"
Event role	"Medical treatment"/"Patient"

Topic Maps

The *topic maps* (2020) is an ISO international standard for representing structured information models. It is a semantic web technology used to express the relationship between abstract concepts and information resources. Therefore, the topic map model can be divided into two separate spaces: (1) Topic space, composed of topics representing real-world concepts; (2) Resource space, composed of resource files which are electronic files such as web pages, texts documents, multimedia files, etc. Connect topics together to form concepts through associative connections and connect resource files through the presence of connections as shown in Fig. 7.22. The components in the topic diagram include: (1) Topics, a machine-processable format that can represent anything about electronics resources or nonelectronic resources (or things in the real world, such as people, places, events, etc.); (2) Association, used to represent the relationship between topics to form concepts; (3) Occurrences used to indicate or refer to related topics resources concept formation.

Ontology

Ontology (Lim et al. 2011) is the most powerful semantic modeling technology among other technologies discussed above. The word *ontology* is borrowed from philosophy. Ontology precisely defines terms related to specific fields, represents the field of knowledge, and standardizes meaning in computer science. According to Gruber (1995): *Ontology is a clear specification for conceptualization.* Ontology consists of a set of words (concepts) usually, taxonomy, relationships, attributes, etc. Therefore, it can model the knowledge domain with stronger semantics than taxonomy and thesaurus.

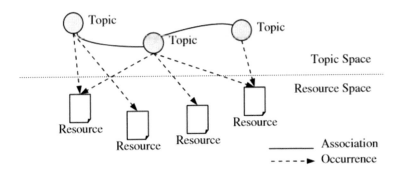

Fig. 7.20 Example of topic maps (Topic Maps 2020; Lim et al. 2011)

Table 7.5 Components of an ontology (Lim et al. 2011)

Component	Description
Classes	Set of concepts that describe objects
Instances	Particular things of objects
Relationships	Associations about meaning among those things
Properties	Property values of those things
Functions	Functions and processes describing those things
Constraints	Description logic and rules describing those things

Table 7.5 lists the components of ontology. As shown, a typical ontology is composed of six ontologies: classes, instances, relationships, properties, functions, and constraints, which correspond to theoretical ontology concepts described in previous sections.

7.8.4   Ontology Languages—RDF

Ontology Languages for Semantic Web

Ontology language is a markup language that can be used to model data semantic architecture in semantic web architecture data layer. The languages that can be used to mark ontology and *semantic web* data semantics include XML, RDF, RDFS, DAML + OIL, and OWL as shown in Fig. 7.21.

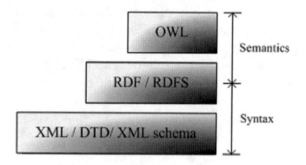

Fig. 7.21 Languages for ontology modeling

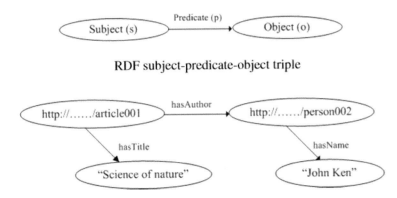

Fig. 7.22 Example of RDF triple relations (RDF 2020; Lim et al. 2011)

Extensible Markup Language (XML)

Extensible Markup Language (XML) is the most basic markup language for data exchange between machines. It is a structured format and can be processed by a machine. XML with a specific DTD or XML schema specifies a grammatical convention but the required data semantics are not defined in XML data, so it is necessary to build an upper markup language on top of XML.

Resource Description Framework (RDF)

RDF (2020) is a language framework by W3C recommendation defined the web-based resource meta-data description.

RDF presents data in subject–predicate–object triple written as P (S, O). This triple notation allows object playing the role of a value, which enables two labeled edges chaining in a graphical visualization as shown in Fig. 7.22. The RDF triples P (S, O) is defined as: hasAuthor(article001, person002), hasTitle(article001, "Science of nature"), hasName(person002, "John Ken"), which can be serialized in RDF/XML syntax as shown in Fig. 7.23.

7.8.5 Ontology Languages—OWL

W3C Web Ontology Language (OWL 2020) is a semantic web language designed to represent rich and complex knowledge about things, groups

```
<owl:Ontology>
  <owl:Class rdf:about="#associateProfessor">
    <owl:disjointWith rdf:resource="#professor"/>
    <owl:disjointWith rdf:resource="#assistantProfessor"/>
  /owl:Class>
</owl:Ontology>
```

Fig. 7.23 Example of RDF

of things, and relationships between things. OWL is a language based on computational logic so computer programs can use the knowledge expressed in OWL. For instance, to verify the consistency of this knowledge or to make implicit knowledge explicit. OWL documents (called ontologies) can be published on the World Wide Web and other OWL ontologies can be referenced or cited from them. OWL is part of W3C semantic web technology stack which includes RDF, RDFS, SPARQL, etc. Compared with RDF and RDFS, OWL provides higher machine readability for web content by adding more vocabulary describing attributes and classes: between classes (e.g. disjoint), cardinality (e.g. exactly one) which cannot be supported by RDF. Therefore, OWL provides more expressive tags for semantic web ontology data. OWL is built on RDF and RDF Schema and uses RDF's XML syntax as shown in Fig. 7.24.

The W3C Web Ontology Working Group has defined OWL as three sub-language: 1. OWL Full; 2. OWL DL; and 3. OWL Lite. Each sub-language is defined for use by specific communities of implementers and users (W3C OWL 2004). OWL current version (also known as OWL 2) was developed by the W3C OWL working group and released in 2009, and the second version was released in 2012. The deliverables that make up OWL 2 specification include a document overview. As an introduction to OWL 2, it describes the relationship between OWL 1 and OWL 2 to provide an entry point for remaining deliverables through the document roadmap. Figure 7.25 shows an OWL2 overview structure.

```
<owl:Ontology>
  <owl:Class rdf:about="#associateProfessor">
    <owl:disjointWith rdf:resource="#professor"/>
    <owl:disjointWith rdf:resource="#assistantProfessor"/>
  /owl:Class>
</owl:Ontology>
```

Fig. 7.24 An example of OWL

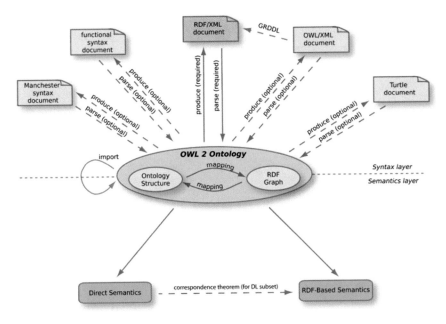

Fig. 7.25 Systematic diagram of NLU (Lim et al. 2020)

7.9 Ontological-Based Search Engine

7.9.1 KnowledgeSeeker

KnowledgeSeeker (Lim et al. 2011) is a comprehensive ontological-based search engine (OSE) system that defines and implements the following components: 1. Ontology modeling (ontology structure); 2. Ontology learning (learning algorithm); 3. ontology generation (format); and 4. Ontology query (operation). Figure 7.26 shows the KnowledgeSeeker system architecture. KnowledgeSeeker can be used to develop various intelligent applications based on ontology by using four defined ontology components. These intelligent applications include knowledge-based information retrieval systems, knowledge mining systems, predicate systems, personalized systems, intelligent agent systems, etc.

Ontology Modeling Module

The ontology modeling module defines a conceptual structure, which is used to represent ontology data (knowledge) in KnowledgeSeeker system. This is a knowledge representation method and knowledge is represented as an ontology graph.

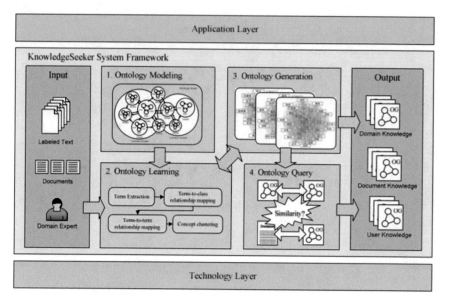

Fig. 7.26 KnowledgeSeeker system framework (Lim et al. 2011)

Ontology Learning Module

The ontology learning module is concerned with acquiring knowledge from text. It defines the conceptualizing method of the knowledge domain. The method is based on a statistical text learner and the conceptualization process is to convert text knowledge into a machine-processable format, that is, the ontology graph defined in the module.

Ontology Generation Module

The ontology generation module formalizes conceptual ontology model into a structure file format. This process uses a text corpus to generate domain ontology (OG) and visualize OG in a graphical format.

Ontology Querying Module

The ontology query module defines how a system operates using ontology graphs. It is an important module that can use KnowledgeSeeker system to develop various smart applications. This module defines operations such as ontology graph matching and querying. These operations make ontology graph data available for various applications development such as text classification systems and text search systems. These applications can also be used to evaluate query performance methods and domain knowledge effectiveness generated in *ontology graphs*.

7.9.2 Ontology Graph

The actual ontology graph (OG) implementation adopts above ontology graph model theory and definition. This section presents conceptual representation and class implementation hierarchy. The ontology graph implementation is used as the basic knowledge representation model in Knowledge-Seeker, which is used for ontology storage, learning, querying, and building ontology-based applications. Figure 7.27 shows an ontology graph conceptual diagram based on term nodes and relationships structure.

According to their complexity in knowledge, *ontology graphs* include four conceptual units (CU) types. Conceptual units (CU), any object (node) in ontology graph can express semantics. All these conceptual units are linked to each other and related through a conceptual relationship (CR) to form

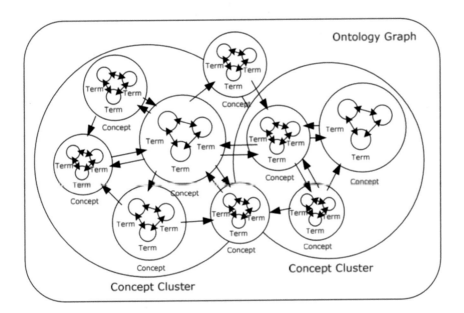

Fig. 7.27 Conceptual structure of ontology graph (Lim et al. 2011)

the entire ontology graph conceptual structure and to model knowledge area (domain).

The definitions of the four conceptual units (CU), their properties, and knowledge levels based on their complexity are described as follows:

(1) *Term (T)*: The smallest conceptual unit extracted in meaningful words which is composed of *meaning* in a human perspective.
(2) *Concept (C)*: Many terms (T) and conceptual relationship (CR) between each other form a concept (C), which is the basic conceptual unit in the concept graph (CG).
(3) *Concept cluster (CC)*: Multiple concepts (C) related to each other form a concept cluster (CC). It groups similar concepts meanings into a tight cluster representing the complex knowledge.
(4) *Ontology graph (OG)*: The largest entire concept unit grouped by concept cluster (CC) is defined as the ontology graph (OG). It represents comprehensive knowledge in a specific field.

7.9.3   Ontology Learning Model

Ontology learning is the process of learning and creating knowledge domains (interests in specific fields such as art, science, entertainment, sports, etc.) in the form of ontology graphs. The creation of ontology graph is considered as knowledge extraction process. We define different levels of knowledge objects in conceptual units (CU), which are necessary for extraction during the learning process. A bottom-up ontology learning method is adopted to extract conceptual units and create ontology graphs. The method identifies and generates conceptual units from the lowest level term (T) to the highest level ontology graph (OG).

The five learning subprocesses start from the bottom are defined as.

(1) *Term extraction*—the most basic process that recognizes meaningful Chinese terms in text documents.
(2) *Term-to-class relationship mapping*—the second process that finds out the relations between terms and classes (domain).
(3) *Term-to-term relationship mapping*—the third process that finds out the relations between all Chinese terms within a class (domain).
(4) *Concept clustering*—the fourth process which further groups (clusters) the Chinese terms within a class (domain) based on their similarities.
(5) *Ontology Graph generation*—the final process that generates a graph-based ontology graph as a knowledge representation for application use.

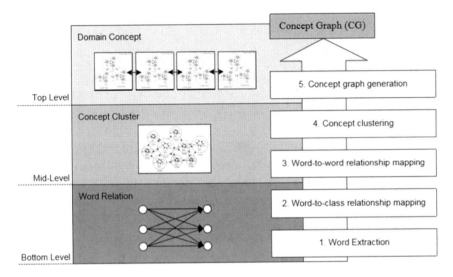

Fig. 7.28 Ontology graph learning process (Lim et al. 2011)

Figure 7.28 shows all subprocesses in the bottom-up approach of ontology graph learning method.

7.10 Applications of OSE in Daily Activities

7.10.1 Intelligent Content Management System

The *intelligent Content Management System* (iCMS) (Lim et al. 2011) is based on OSE technology. It provides solutions for different content providers, i.e. publishers, media, new institutions, libraries to centrally manage, organize, search, data mine, archive, and retrieve digital assets such as news articles, photos, videos, audio clips iCMS database. By integrating iCMS with the ontology knowledge search system, all digital content can be enhanced and organized through ontology-based knowledge. Digital content can be retrieved and distributed through different channels and mobile applications. iCMS KnowledgeSeeker is an ontology system for managing and organizing all digital content in iCMS through ontology methods. iCMS Knowledge-Seeker is composed of the content database cluster, ontology index database, and ontology search engine. The iCMS KnowledgeSeeker search engine uses ontology methods to analyze Chinese text content such as news articles and uses semantic web concept to semantically organize information. iCMS

KnowledgeSeeker also uses ontology method to identify the article subject (text classification process).

iCMS KnowledgeSeeker consists of three components:

1. Ontology and content index—it stores all ontology information and all analysis information about all iCMS content including ontology-based index.
2. Ontological search engine—it combines content analysis, content indexing, index search, and user response to search results.
3. iCMS database cluster—it stores all content files' original sources including articles, audio, video, images, electronic publication data, etc.

The processing flow between these components is shown in Fig. 7.29.

7.10.2 Intelligent News Retrieval and Ontological Search Engine

OSE News (Lim et al. 2011) is a web platform for reading Chinese RSS news feeds. It has integrated intelligent agent technology (IAT) (Lee and Loia 2007), which enable browsers to read RSS news articles and also-decompose all articles and provide other related articles to users automatically. In addition, it also allows each user to create their own personalized news categories. The main functions and features of OSE News include: an intelligent agent system for network news collection; an ontology system for domain knowledge modeling; a 5D ontology system for news semantic analysis and personalized categories. Figure 7.30 shows OSE News system architecture. It integrates an ontology system for analyzing news content and identifying news topics automatically. The core ontology knowledge also enhances the news search engine so that it can provide users with more accurate and relevant results. The intelligent self-learning function can also provide news personalization for each registered user. Each user receives their personalized content based on reading habits and interests. They can enter areas of interest into the system when reading news through website, or let the system learn. This allows users to receive content they are most interested in and filter out most content that is not of interest.

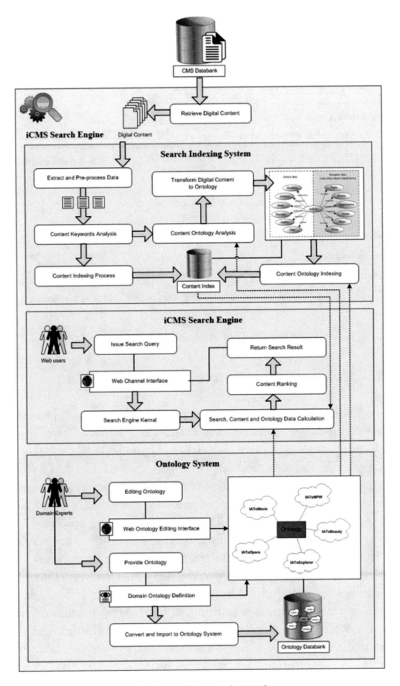

Fig. 7.29 iCMS using OSE technology (Lim et al. 2011).

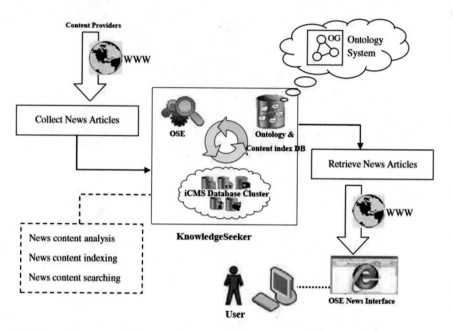

Fig. 7.30 OSE news architecture (Lim et al. 2011)

7.10.3 Intelligent Web Ontology Learning System

The *collaborative learning method* is mainly divided into two processes: (1) *Content-based ontology learning process;* and (2) *User-based ontology personalized learning process* (Lim et al. 2011). Figure 7.31 shows the system architecture of the learning method. This figure shows a data flow overview of the entire learning system. Basically, the two learning processes are processed separately, because learning result of creating the content-based learning process is taken as basic input of the user-based learning process. Ontology extraction and verification are performed on the user side. They modify and refine ontology and input it into the content-based learning process to create a complete learning cycle to improve ontology effectiveness. This content-based ontology learning process includes four main steps: 1. Text analysis; 2. Concept selection; 3. Ontology learning; 4. Ontology verification. Ontology learning outcomes—ontology graph (OG) is defined in this learning process. The text analysis process of Chinese documents in network channels requires common Chinese terms and special terms listing for each network channel domain. Extract common Chinese terminology from electronic dictionaries (such as HowNet) which contain more than 50,000 different Chinese characters used as an initial terms list for the text analysis process. The special web channel domain terms such as named entities, product brands, product

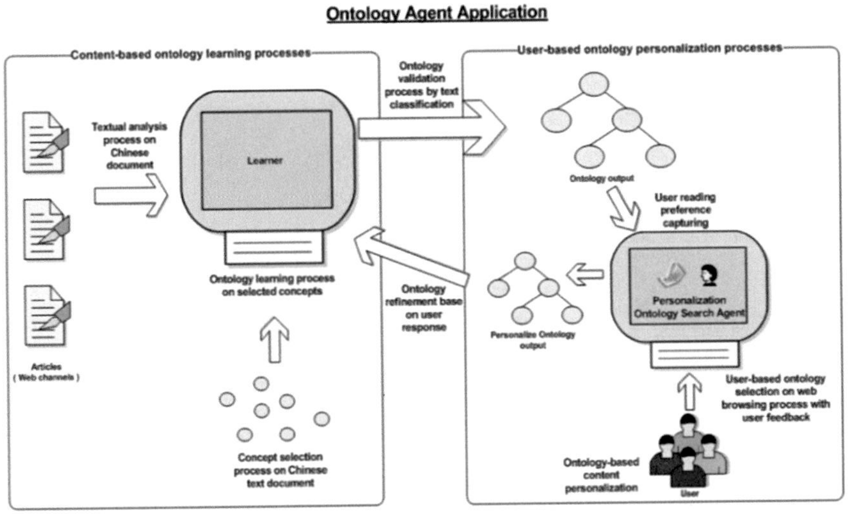

Ontology Agent Application

Fig. 7.31 OSE on ontology learning system (Lim et al. 2011)

models, etc. are manually defined. The combination of special terms and initial term list is the only predefined knowledge in the ontology learning process. Then, the maximum matching algorithm is applied to the term list and the web channel document to extract the candidate term (CT) list so that each term in the list exists at least once in all web channels.

7.11 Case Study: Language Learning Robots Using NLP

Up to this chapter, we have learnt the basic AI concept and the five AI core technology that include machine learning, data mining, computer vision, natural language processing, and ontology search engine.

Technically speaking, we have sufficient knowledge to build *ontology tree* or *ontology graph* on knowledge related to AI.

Suppose you are assigned to form a team to study and construct an ontological-based learning system on AI:

(1) Based on various ontology construction technologies learnt in this chapter, build an ontology structure (can be ontology tree, graph, or map) to relate all concepts, ideas, and technology related to AI learnt in this course (up to this chapter).

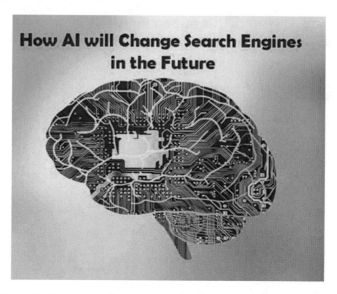

Fig. 7.32 How AI change search engine in the future

(2) In Chap. 6, we have learnt all related concepts and technologies related to NLP (Natural Language Processing). Discuss and explain how we can integrate OSE-based AI learning system into NLP to provide an NLP-based AI learning system.

(3) As a group discussion topic, discuss is it possible for a computer to generate *new knowledge* based on all AI technology learnt so far? (Fig. 7.32).

7.12 Conclusion

In this chapter, we discuss the last but also one of the foremost AI technology—*ontological search engine (OSE)*. We start with a traditional search engine overview mainly focusing on keyword search and web crawler and indexing technology. However, owing to the exponential web resources and information increases over the Internet on a daily basis, current search engine technology cannot cope with the increasing demand in terms of search results quality, a new age of search engine emerged—*Ontological-based Search Engine*.

Ontology is not in fact a new idea in human history. It is the core of human knowledge actually and the way we percept and understand concepts, meanings, and ideas. In the second part of this chapter, we explore various latest ontological search engine technology and discuss how AI scientists

Fig. 7.33 World of ontologies

design and implement ontology frameworks, systems, and related ontology languages such as OWL and OWL2. We also discuss some potential OSE technology applications that include OSE on automatic new retrieval, intelligent content management system (iCMS), and intelligent web ontology learning systems. OSE technology development is in fact the beginning, hopefully more new technology and applications will appear in coming decades and can surely reshape our world and our ways to access information and knowledge (Fig. 7.33).

References

Abbas, J. (2010). *Structures for organizing knowledge: Exploring taxonomies, ontologies, and other schema*. Neal-Schuman Publishers.

Allemang, D., & Hendler, J. (2011). *Semantic web for the working ontologist: Effective modeling in RDFS and OWL* (2nd ed.). Morgan Kaufmann

Andrew, M. (2018). *How to Google image search: Discover how to google reverse image search and find out more information about any picture online* (Kindle edition). Amazon.com.

Aristotle. (2014). *The categories* (E. M. Edghill, Trans.). CreateSpace Independent Publishing Platform.

Berners-Lee, T. (2000). *Weaving the web: The original design and ultimate destiny of the World Wide Web.* Harper Business.

Brachman, R., & Levesque, H. (2004). *Knowledge representation and reasoning* (The Morgan Kaufmann Series in Artificial Intelligence). Morgan Kaufmann.

Büttcher, S., et al. (2010). *Information retrieval: Implementing and evaluating search engines* (The MIT Press). MIT Press.

Cognitive Science Lab. (2009). *WordNet 3* (largest English dictionary and thesaurus, Kindle edition). OSNOVA.

Croft, B., et al. (2009). *Search engines: Information retrieval in practice.* Pearson.

Denning, P. J., & Tedre, M. (2019). *Computational thinking* (MIT Press Essential Knowledge series). The MIT Press.

Dong, Z., & Dong, Q. (2006). *HowNet and the computation of meaning.* Hackensack, NJ: World Scientific.

Effingham, N. (2013) *An Introduction to Ontology.* Polity.

Gruber, T. R. (1995). (1995) Toward principles for the design of ontologies used for knowledge sharing. *International Journal of Human-Computer Studies, 43*(5), 907–928.

Guarino, N. (1995). Formal ontology conceptual analysis and knowledge representation. *International Journal of Human-Computer Studies, 43,* 625–640.

Gunjan, V. K., & Subramanyam, A. (2014). *Search engines-how do they work?: Crawlers & SEO.* Scholars' Press.

Haider, J., & Sundin, O. (2019). *Invisible search and online search engines: The ubiquity of search in everyday life.* Routledge.

HowNet. (2020). Retrieved May 9, 2020, from Official site: https://www.keenage.com/html/e_index.html.

Kaye, L. J. (2019). *Kant's transcendental deduction of the categories: Unity, representation, and apperception.* Lexington Books.

IEEE. (2020). *IEEE Upper Ontology Standard.* Retrieved June 10, 2020, from https://site.ieee.org/pes-mas/upper-ontology/.

Lee, R. S. T., & Loia, V. (2007). *Computational intelligence for agent-based systems.* New York, Berlin: Springer.

Lee, J., et al. (2019). *An effective approach to enhancing a focused crawler using google. The Journal of Supercomputing.* https://doi.org/10.1007/s11227-019-02787-9

Levene, M. (2011). *An introduction to search engines and web navigation* (2nd ed.). Wiley

Lim, E. H. Y., Liu, J. N. K., & Lee, R. S. T. (2011). *Knowledge seeker: Ontology modelling for information search and management : A compendium.* Berlin: Springer.

Mascardi, V., et al. (2010). Automatic ontology matching via upper ontologies: A systematic evaluation. *IEEE Transactions on Knowledge and Data Engineering, 22*(5), 609–623.

Miller, G. A. (1998). *WordNet: An electronic lexical database.* MIT Press.

OpenCyc. (2020). Retrieved May 9, 2020, from Official site: https://slor.sourceforge.net/e_ocyc.htm.

OWL. (2020). Retrieved May 9, 2020, from Official site: https://www.w3.org/OWL/.

Rastogi, S. R., et al. (2013). Search engine techniques: A review. *MIT International Journal of Computer Science & Information Technology, 3*(2), 53–57.

RDF. (2020). Retrieved May 9, 2020, from Official site: https://www.w3.org/RDF/.

Seltzer, R., et al. (1997). *The AltaVista search revolution: How to find anything on the Internet.* Mcgraw-Hill Osborne Media.

Sheth, A. (2013). *semantic web: Ontology and knowledge base enabled tools, services, and applications.* IGI Global.

Simiraglia, R. (2015). *Domain analysis for knowledge organization: Tools for ontology extraction.* Chandos Publishing.

SUMO Ontology. (2020). *Suggested Upper Merged Ontology.* Retrieved June 10, 2020, from https://www.adampease.org/OP/.

Topic Maps. (2020). Retrieved May 9, 2020, from Official site: https://www.topicmaps.org/.

W3C. (2020). *W3C Semantic Web.* Retrieved May 9, 2020, from Official site: https://www.w3.org/standards/semanticweb/.

Willaschek, M. (2018). *Kant on the sources of metaphysics: The dialectic of pure reason.* Cambridge University Press.

WordNet. (2020). *WordNet.* Retrieved May 9, 2020, from Official site: https://wordnet.princeton.edu/.

Part III

AI Applications

8

Intelligent Agents and Software Robots

There are lots of examples of routine, middle-skilled jobs that involve relatively structured tasks, and those are the jobs that are being eliminated the fastest. Those kinds of jobs are easier for our friends in the artificial intelligence community to design robots to handle them. They could be software robots; they could be physical robots.
Prof. Erik Brynjolfsson (Educator, born 1962)

Abstract This chapter studies the first type of AI applications—intelligent agents. Intelligent agent technology (IAT) can be considered as AI exemplification in the volatile and mobile environment. IAT provides a truly robust and autonomous software agents implementation in terms of agent connectivity from existing client–server technology evolution that has been used for over half a century. It can also be regarded as the soul of robot, or software robot quoted by many AI scientists in terms of AI perspective. The chapter covers its basic requirements and explores different varieties of frameworks followed by several major applications that include agent shoppers, agent negotiator, agent weatherman, and agent traders. Lastly, we explore the threats and challenges of IAT.

Anyone watched sci-fi action movie series *Matrix* directed by the Wachowskis in the late 1990s knew who Neo was. A hacker who discovered the *truth* of our reality that we all lived in a *construct* so-called *Matrix*, a simulation system designed by AI machine to control us in the future world. We all

© The Editor(s) (if applicable) and The Author(s), under exclusive license
to Springer Nature Singapore Pte Ltd. 2020
R. S. T. Lee, *Artificial Intelligence in Daily Life*,
https://doi.org/10.1007/978-981-15-7695-9_8

are trapped inside the construct as *prisoners*. All our senses and feelings are no more than simulations generated by the construct to deceive us living in *reality*. The whole world was being destructed in the late twentieth century. Humans are trapped by the AI machine as energy resources and living with shared consciousness in a simulation world (Irwin 2005).

The most striking fact to AI scientists working in the area of intelligent agents inspired by this movie was that not only the notion of *simulation theory* was a niche and popular topic at that time, but also the character *Agent Smith*—the villain who fought against Neo and his team tirelessly. He was not a person but a software program object so-called *intelligent agent (IA)* in AI, also known as *software robot*. This was different from the traditional program that it lives in a computer system, intelligent agent can exist independently (or what we called *autonomously*) throughout the network to execute his designated duty and operation. It can also acquire new knowledge, skills on demand, and *clone* himself into multiple copies to finish designated tasks.

Does an intelligent agent really exist or only a science fiction?

We will explore this fascinating technology based on R&D works on how it can reshape our daily activities and various kind of services in a new world of an intelligent city.

8.1   Intelligent Agents—The Soul of Robotics

Many AI scientists define *intelligent agents* as the *soul of robotics*. What is the *soul*? Does it really exist? or only a fantasy thought?

Since the beginning of human civilization, philosophers, religious figures, scholars, and even neuroscientists have been searching for an answer to this ultimate question. One of the most distinctive and widely accepted theories is the so-called *Mind–Body Dualism*.

The *mind–body dualism* theory was first discussed by Plato (427–347 BC) in his *Phaedo* (Plato 1999) which interpreted the nature of existence as a *world of two realms*: (1) The realm of matter properties which is by its nature – our *immortal soul;* and (2) The realm of matter itself which is by its nature—change and decay with time—our *mortal body*. Aristotle (384–322 BC) shared Plato's view of immortal souls and further elaborated a hierarchical arrangement, corresponding to distinctive functions of plants, animals, and human: a nutritive soul of growth and metabolism that all three share; a perceptive soul of pain, pleasure, and desire that only human and other animals share; and the faculty of reason that is unique to human only (Jaworski 2011; Nyvlt 2011).

The modern problem of the relationship of mind to body stems from the thought of René Descartes (1596–1650), a seventeenth-century French philosopher, mathematician, and scientist who gave dualism its classical formulation. Beginning from his famous *Cogito, ergo sum (Latin: "I think, therefore I am")* (Descartes 1998), he developed a mind theory as an immaterial, non-extended substance that engages in various activities such as rational thought, imagining, feeling, and willing. Matter or extended substance conforms to the laws of physics in mechanistic fashion with the important exception of the human body, which he believed is affected causally by human mind and produces certain mental events causally. Figure 8.1 shows Descartes' illustration of dualism. Inputs are passed on by sensory organs to the epiphysis in the brain and from there to the immaterial spirit. He clearly identified the mind with consciousness and self-awareness and distinguished this from the brain as the seat of intelligence. Hence, he was the first to formulate the mind–body problem in the form in which it exists today.

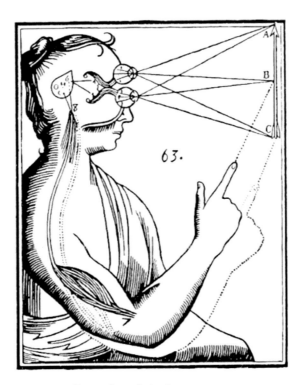

Fig. 8.1 René Descartes's illustration of dualism

From the modern computer science perspective, we can regard the computer system as a kind of *mind–body dualism* in which computer machine is the *body*. It is driven by its *soul*—the operating system (OS) and related software programs that can exist and operate autonomously with its *host machine*. Extended to modern robotics, the *soul* of the robot can regard as software and OS that contain all knowledge, memories, experiences, and decision-makings to *drive* and *control* the robot every single moment; that can exist independently with the robot itself— the *intelligent agents*, the soul of robotics (Lee 2006; Lee and Loia 2007).

8.2   What Are Intelligent Agents?

Strictly speaking, an intelligent agent (IA) is a combination of two concepts: *AI* and *mobile agent*. The concept of intelligence has already been revealed at the beginning of this book. Therefore, let us focus on the agent definition. The word agent is simple. It has three basic meanings in *Oxford English Dictionary*: (*1*) *a person who acts or manages affairs for another person or another person; (2) a person has achieved something in order to achieve achievement; (3) in science, agent refers to the substances that produce effects* (Lee 2006; Lee and Loia 2007; Kolodziej et al. 2015).

We can consider IA as a computer *agent* (or *program*) that can help us manage the other business intelligently. Given that we have a certain understanding of what intelligence and AI are, let's take a look at what IA is. A chapter in the popular textbook *AI: A Modern Approach* by Prof. Russell and Prof. Norvig (2020) is about IA. Although there is no formal IA definition, the rational agent is defined as: *For each possible perception sequence, supported by the evidence provided by the perception sequence and any built-in evidence, the rational agent should choose one that is expected to maximize its action performance.* This book also defines an agent as: *An agent can be seen as anything that perceives its environment through sensors and acts on the environment through actuators.*

Another book on *intelligent software agents* (Murch and Johnson 1998) written by Prof. Richard Murch and Prof. Tony Johnson divides agents into four types: *intelligent, learning, mobile, and believable agents.* There is no direct definition for intelligent agents. *The only paragraph mentioned about the definition of intelligent agents is that: These (intelligent agents) are very broad categories of agents that can go beyond the definitions described previously. This is the largest area of research, and software developers show the greatest business interest.*

Fig. 8.2 Intelligent agents and robotics (Tuchong 2020a)

From the author's point of view (Lee 2006), an intelligent agent (IA) is the *exemplification of human intelligence in a device.* The agent's intelligence consists of possessing knowledge with three levels: *derived knowledge, stimulated knowledge, and intuitive knowledge, and the manipulation of knowledge—the thoughts or thinking consists of three levels: (logical thinking, lateral thinking, and intuition). This device (IA) can exist in the form of a system, a software program, a program object, or even robots* (Fig. 8.2).

8.3   Basic Requirements of Intelligent Agents

There are ten basic requirements of a typical intelligent agent should (or might) possess (Lee 2006) in addition to the fundamental intelligence requirements for ease of implementation.

1. *Autonomous*—exist and survive independently.
2. *Mobile*—traverse and navigate freely, e.g. over the Internet to complete the task.
3. *Proactive*—express requests and/or implement preventive measures before adverse events occur. For example, before the other party provides

a further or fresh offer, the active negotiating agent provides other options during product negotiation.

4. *Reactive*—provide a rational response to external stimuli and inquiries.
5. *Adaptability*—adapt to the external environment. For example, a highly adaptable shopping agent can generate its clones dynamically based on the number of electronic shopping malls it needs to visit. In extreme cases, it should calculate the optimal number of clones that need to be generated dynamically based on the number of shopping sites, current network traffic, and loading costs.
6. *Robustness*—robust to the external environment. For example, a highly robust shopping agent can perform tasks in various environments and platforms, so that it can be called and travel freely in Web and wireless environments especially when the environment has limited resources and bandwidth.
7. *Co-operative*—sociable and co-operate with foreign agents. For the so-called multi-agent system, this is one of the IA main requirements and characteristics.
8. *Learning*—can self-improve its intelligence through learning. Another IA major characteristic is not only the possession of *priori knowledge*, but also its learning ability using supervised, unsupervised, or reinforcement learnings. For example, a smart auction agent should have the ability to learn how to bid on auction sites so that bidding can be conducted wisely in the future.
9. *Task-oriented*—IA should have some predefined tasks. Its purpose is to use its own intelligence to try its best to accomplish these tasks.
10. *Goal-oriented*—based on the goals achieved. A good IA should not do anything that has nothing to do with the goal. For example, goal-driven shopping agents will not (and should not) visit any non-shopping sites to search for products.

8.4   Variety of Intelligent Agents

According to the above IAs requirements and characteristics, the majority of agent developers focus on building one (or several) capabilities for their agents leading to a variety of (what may be called) functional agents in the market. They include.

(1) *Multi-agents*—focus on multi-agent communication and knowledge exchange (Ferber 1999; Shen 2001).

(2) *Internet agents*—focus on software agents' design and implementation over the Internet platform (Omicini et al. 2001).

(3) *Mobile agents*—focus on mobility over distributed networks, or possibly over the Internet or some other propriety networks (Rothermel and Popescu-Zeletin 1997; Baumann 2000).

(4) *Goal-based agents*—expand further on capability by using *goal* information. Goal information describes situations that are desirable. This allows the agent to choose the one among multiple possibilities to reach a goal state. Search and planning are devoted AI subfields to find action sequences to achieve goals.

(5) *Learning agents*—focus on IAs machine learning functionality (Furukawa et al. 1999). Typical examples are PA (Personal Assistant) agents that target interactive learning of their owners' *habits*.

(6) *Adaptive agents*—agents that focus on the adaptiveness to the external environment in the sense that an adaptive agent can modify itself, in terms of its thinking, knowledge, and action/reaction activities according to changes in the external environment (Alonso et al. 2003). Most adaptive agents also possess certain learning abilities.

Figure 8.3 shows an IAs variety overview.

From the author's point of view, different kinds of agents (in the market) should also be categorized as IAs. The only difference between them is the degree of intelligence and the focus on the agent's capabilities. For instance,

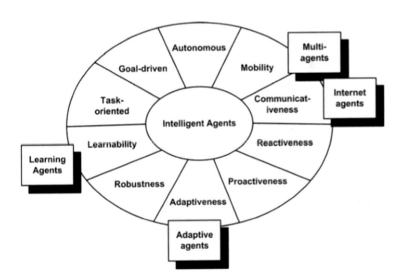

Fig. 8.3 An overview of intelligent agent variety (Lee 2006)

multi-agents are not only restricted to possess communication and mobile abilities but also can be intelligent enough to learn and proactive naturally.

8.5 General-Purpose Intelligent Agent (GIA)

Conceptually, our daily psychological activities are just the so-called *mind–body* interactions mentioned at the beginning of this chapter. The main IA concept is to embody human intellectual activities in certain devices. In other words, the most direct and appropriate IA method implementation model should be this interaction model direct embodiment, the so-called *General IA (GIA) structural mode*l. Figure 8.4 shows a GIA system structural model (Lee 2006).

This GIA structural model is called a general model because it has all common knowledge and IA technical knowledge to handle all daily activities. As shown, the GIA model only applies psychological activities of the human body and mind directly to the IA system, with body and mind interaction described by the interface between *perception module* (PM) and *knowledge thinking module* (KTM) and *reaction module* (RM). PM corresponds to the entire perceptual stimulus (or query) in the environment; KTM corresponds to knowledge storage and inference in the agent's mind; RM corresponds to all actions and responses taken by the agent.

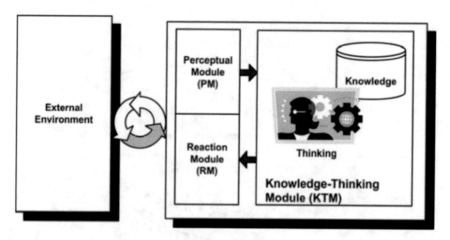

Fig. 8.4 GIA structural model (Lee 2006)

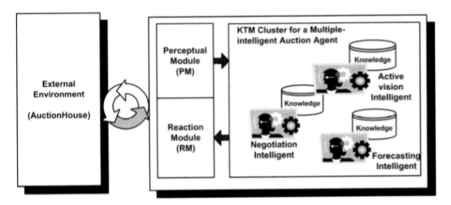

Fig. 8.5 TIA structural model for a multiple intelligent auction agent (MAA) (Lee 2006)

8.6 Task-Oriented Intelligent Agent (TIA)

The GIA model is an ideal case at least it is not easy to use today's technology to implement a general agent with all possible common sense and technical knowledge. The IA can be developed in a so-called *intelligent-on-demand (IoD)* mode to solve this problem. That is based on a similar architecture, KTM is now divided into different functional modules. Each submodule of KTM will represent different intelligent modules required by the agent to handle specific tasks. Therefore, the agent will be developed as an integrated smart device based on *task-oriented* IA (TIA) development model (Lee 2006). Figure 8.5 shows an intelligent auction agent with three different *plug-in intelligences*: active visual intelligence, e.g. for product visualization; negotiation intelligence, e.g. for product bargaining; and prediction intelligence, e.g. for auction trend prediction. Today, this IoD solution provides an inevitable advantage in IA development. Most importantly, the agent development platform can not only serve as a general place for IA to initiate and implement, but also as a *smart store* to provide these smart modules on demand.

8.7 Intelligent Agent Framework

To corporate with smart city future development and 5G integration, AI and *intelligent agent technology (IAT)*, an innovative intelligent agent development platform is constructed at United International College (UIC) for the design and implementation of the future 5G-based smart city with related intelligent agent-based applications. Figure 8.6 illustrates the UIC 5G Intelligent Agent

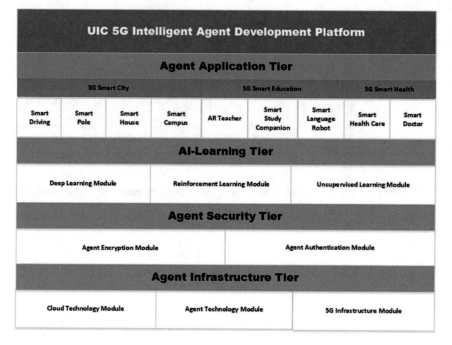

Fig. 8.6 UIC 5G intelligent agent framework

Framework system infrastructure.

The *UIC 5G intelligent agent system framework* provides a four-tier implementation that includes.

> *Application tier*—Based on the infrastructure provided by *agent infrastructure tier, agent encryption* and *authentication function modules* provided by *agent security tier,* together with AI functions provided by *AI-learning tier, application tier* implements UIC 5G intelligent agent solution in three major aspects: (1) *5G smart city* includes *smart driving, smart pole, smart house, and smart campus;* (2) *5G smart education* includes *AR teacher, smart study companion, smart language robot;* (3) *5G smart health includes smart health care and smart doctor.*
>
> *AI-learning tier*—With the implementation of three kinds of AI-learning technology: *deep learning, reinforcement learning, and unsupervised learning,* AI-learning tier provides all necessary functions and APIs for implementation of UIC 5G intelligent agent applications in the *application tier.*

Agent security tier—Implementation of agent-based APIs for multi-agent messages/data encryption and authentication for every UIC 5G Intelligent Agent services.

Agent infrastructure tier—Provides basic infrastructure, functions, and APIs in three major areas: *cloud technology, agent technology, and 5G infrastructure modules* for UIC 5G intelligent agent solutions development.

8.8   Intelligent Agent Applications

8.8.1   Agent Shoppers

Agent shoppers can provide integrated, agent-based smart solutions for smart shopping through any mobile device. It is *intelligent agents* integration with the following technologies: (1) mobile agent technology based on FIPA (2020) agent standards (2) Java servlet technology for manipulating server-side operations in agent computers, and (3) intelligent FShopper based on fuzzy-neural smart shopping operations. Figure 8.7 illustrates agent shopper overall system framework based on agent shopping using intelligent agent technology in different online stores. It also demonstrates two cases of smart agent shopping: (1) fuzzy Internet shopping via a web browser, and (2) fuzzy mobile shopping using a mobile phone such as an iPhone or Android device.

In other words, any agent-based online store can run in this framework as long as its proxy server conforms to FIPA (Intelligent Physical Agent Foundation) standard. More importantly, both Web-based electronic shopping and mobile shopping can run simultaneously under this infrastructure.

The agent shopper system framework consists of six main modules as follows:

- Customer requirement definition (CRD);
- Requirement fuzzification scheme (RFS);
- Fuzzy agent negotiation scheme (FANS);
- Fuzzy product selection scheme (FPSS);
- Product defuzzification scheme (PDS); and
- Product evaluation scheme (PES).

Agent shoppers provide a brand-new intelligent shopping vision compared with traditional online shopping. Once agent shoppers have collected user requirements and preferences, they can clone themselves into multiple avatars for online shopping. They can carry out price comparison and even product

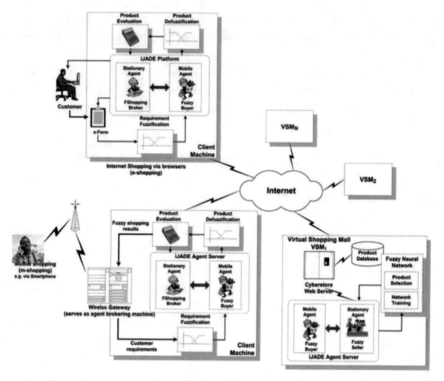

Fig. 8.7 Agent shopper system framework (Lee 2006)

negotiation; and communicate with multiple online stores sales agents world-wide simultaneously.

8.8.2 Agent Negotiators

We often negotiate with the store owner about products with different attributes such as price, quality, delivery date, etc. of interest in a typical e-shopping scenario. All these negotiation attributes are highly dynamic and in terms of their significance, this is a complex and intellectual task. Throughout the bargaining process, a fixed algorithm is used to model all changes in these negotiation attributes. In these cases, we are dealing with varying degrees of ambiguity, which can be manipulated by AI technologies such as fuzzy systems effectively. Through machine learning technologies integration such as neural networks, an intelligent product selection, consultation, and negotiation system has been formed. *Agent Negotiator*, an innovative multi-agent-based intelligent negotiation system that utilizes fuzzy negotiation strategies and utility function theory integration to achieve dynamic, adaptive, and fully automatic multi-agent-based product negotiation.

From a system architecture perspective, the *Agent Negotiator* shown in Fig. 8.8 consists of four main entities (Lee 2006):

(1) *Buyer agent (BA)*—BA is to capture user's needs and search for the most suitable products for users. Another function is to return the seller's negotiation model to the buyer negotiator (BN). BA is a mobile agent actually. Once the user's order is received, it will make a copy and send it to other remote shopping malls.

(2) *Buyer negotiator (BN)*—BN is responsible for capturing user's negotiation requirements, analyzing the negotiation rules, and dispatching them to the remote agent competition. BN is a mobile agent and contains five main modules.

(3) *Seller agent (SA)*—SA is the salesman agent of a remote shopping mall. It is a stationary agent and always participates in the competition with other seller agents. SA will provide purchasing agents with product and negotiation information.

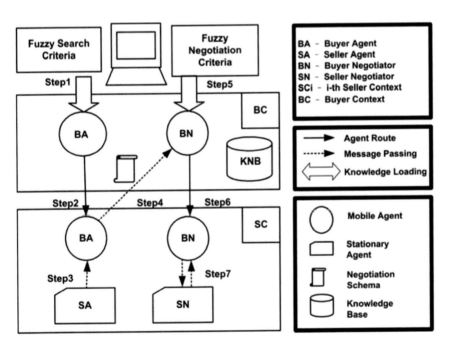

Fig. 8.8 System architecture of Agent Negotiator (Lee 2006)

(4) *Seller negotiator (SN)*—SN is also a stationary agent waiting for BN request and negotiating with the buyer. SN is to exchange knowledge between the two parties and optimize transaction results.

8.8.3   Agent Weatherman

Weather forecasting has been one of the most challenging problems in the world for centuries, not only because of its practical value in meteorology, but also as a time-series forecasting problem that is usually unbiased in scientific research. This section introduces *Agent Weatherman*—a weather forecast agent that uses an intelligent fuzzy-neural system based on multi-agents, which automatically collects and data-mine weather information and performs time-series weather forecasting (operated by a fuzzy-neural network model) with weather information provided by various weather stations. Agent Weatherman provides an innovative, intelligent agent-based prediction system that can extract and analyze weather information. The system provides data provided by multiple weather stations in Hong Kong.

Agent Weatherman consists of five main components (Lee 2006):

- User requirement definition scheme (URDS),
- Data collection scheme (DCS),
- Variable selection and transformation scheme (VSTS),
- Fuzzy–neuro training and prediction scheme (FNTPS), and
- Weather reporting scheme (WRS).

Figure 8.9 illustrates an Agent Weatherman Schematic diagram.

URDS involves the collection of user needs, including the selection of one or more forecast metrics, i.e. temperature, rainfall, humidity, etc. forecast range i.e. the next or N-day forecast and other parameters such as regional weather forecast or global weather forecast. This information is collected through the *Agent Weather Reporter*, which is located on the client computer and is a stationary agent used to collect user needs and negotiate and dispatch the *Agent Weather Messenger* to process WRS. In the DCS scheme, each weather messenger (mobile agent) visits different weather stations to collect weather information and then centralizes it in the main weather center for further processing. Once all *Weather Messenger* is centralized in the central station, they will exchange and integrate the weather information they collected and reorganized (involving selection, information grouping, and conversion of weather elements) so that they can be used by the *Weather Forecaster Agent* which conducts system training and testing. After

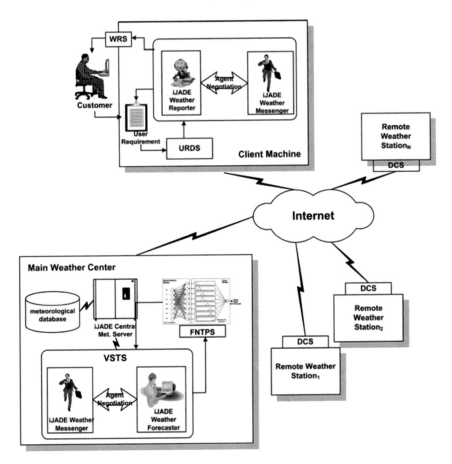

Fig. 8.9 System architecture of agent weatherman (Lee 2006)

collecting and preprocessing all relevant weather information, the *Weather Forecaster* (stationary agent located at the central station) begins appropriate network training and forecasting based on using fuzzy-neural network.

8.8.4 COSMOS Traders

Financial engineering ranging from financial signals study to financial prediction modeling is one of the most stimulating topics for both academia and financial community over the years. Not only because of its importance in terms of financial and commercial values, but more it poses a real challenge to worldwide researchers and quants vitally owing to its highly chaotic and almost unpredictable nature. *Cosmos Traders* (Lee 2019, 2020) devises an innovative *Chaotic Oscillatory Multi-agent-based Neuro-computing System*

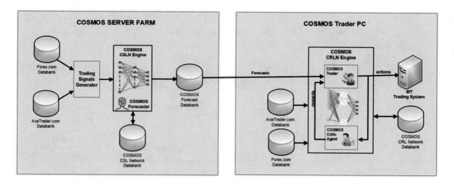

Fig. 8.10 System architecture of COSMOS Traders (Lee 2019)

(a.k.a. COSMOS) for worldwide financial prediction and intelligent trading. With the adoption of the author's theoretical works on Lee-oscillator with profound transient-chaotic property, COSMOS effectively integrates chaotic neural oscillator technology into: (1) COSMOS Forecaster—Chaotic FFBP-based Time-series Supervised learning agent for worldwide financial forecast and; (2) COSMOS Trader—Chaotic RBF-based Actor-Critic Reinforcement learning agents for trading strategies optimization. Figure 8.10 illustrates Cosmos Trader system architecture.

COSMOS not only provides a rapid reinforcement learning and forecast solution, more prominently it resolves successfully massive data over-training and deadlock problems usually imposed by traditional recurrent neural networks and RBF networks using classical sigmoid or Gaussian-based activation functions. From the application perspective, real-time and historical data of worldwide 129 financial products provided by Forex.com (2020) (major online Forex trading platform) and AvaTrade.com (2020) (the biggest cryptocurrency trading platform) are adopted for COSMOS system implementation. They include major cryptocurrencies (9), worldwide forex (84), major commodities (19), worldwide financial indices (17). From an implementation perspective, COSMOS is integrated with 2048-trading day time-series financial data and 39 major financial signals as input signals for the real-time prediction and intelligent agent trading of 129 worldwide financial products which consists of: 9 major cryptocurrencies, 84 forex, 19 major commodities, and 17 worldwide financial indices. In terms of system performance, the past 500-day average daily forecast performance of COSMOS attained less 1% forecast percentage errors and with promising results of 8–13% monthly average returns.

8.9   Agent Technology—Threats and Challenges

Future intelligent agents might be equipped with more powerful intelligence in view of technology advance of computational speed and memory capacity. It is a challenge for us to develop agents to meet these ever-increasing needs in solving more complex, fuzzy, and highly dynamic problems. Another challenge is whether we can build general-purpose and multifunctional agents with human-like common sense and knowledge finally. However, there are several risks or threats we must manage with the increasing number of agent applications.

The most critical one is the security which has two aspects: (1) the risk or threat of intruder agents; and (2) the potential risk or threat of personal loss (and important) information from our own agents to the outside world.

An agent marketplace without any security (or housekeeping) control is alike opening house door and inviting anyone to enter, even a firewall may not fully protect our assets (our information). Standardization might be a possible solution, but it must be reviewed from time to time, especially for those that conform to the IEEE FIPA standard, one of the most popular international standards for agent application design and implementation (Fig. 8.11).

Fig. 8.11 Agents threats and challenges (Tuchong 2020b)

8.10   Case Study: Shopping and Bargaining Agents

In this chapter, we have learnt the intelligent agent technology's basic concept and theory. There are different kinds of intelligent agents studied that can be applied to our daily activities. In this case study, each group is assigned to design a new age of intelligent online shopping and bargaining agents so-called *smart shopper*.

In addition to the multi-agent-based shopping service previous discussed in this chapter, smart shopper can also have the ability to bargain with sale agents (or promoters) in every cyberstore it visited, and come back with the best price and quality of product to user.

(1) List out all functions and services smart shopper needs to provide.
(2) Discuss and explain what kinds of AI technology is required to implement all these functions and services.
(3) What are the major challenge and difficulties to implement this intelligent agent application? And discuss how to overcome these challenges and difficulties.
(4) A lot of customers have changed their habits from physical shopping to online shopping nowadays. Many physical stores try to promote *Online-to-Offline commerce (so-called O2O)* in many cities. The main purpose of O2O Commerce is a business strategy that draws potential customers from online channels to make purchases in physical stores. Discuss how smart shoppers can also be applied to promote O2O as a value-added service? (Fig. 8.12).

Fig. 8.12 Shopping and bargaining agents (Tuchong 2020c)

8.11   Conclusion

We discuss the first type of AI applications—*intelligent agents* in this chapter. Intelligent agent technology (IAT) can in fact be considered as AI exemplification in a volatile and mobile environment.

IAT provides a truly robust and autonomous software agents' implementation in terms of agent connectivity, which can be regarded as an existing client–server technology evolution that has been used for over half a century.

It can be regarded as the *soul* of robot, or *software robot* quoted by many AI scientists in terms of AI perspective.

IAT progress and development is hindered by mobile devices' capability in the past decades. But thanks for the exponential advance in mobile technology in the past 10–15 years, CPU and memory capabilities of the latest mobile devices such as iPhone or Android devices are even more powerful than desktop computers nowadays.

Together with the latest 5G Technology for the provision of greater bandwidth and download speed, seamless positioning, and low latency; a new wave of Smart City applications, ranging from auto-driving to smart pole, to smart education and smart health applications, AI integration, 5G, Data Mining, Computer Vision, Natural Language Processing, Ontological Search Engine, and Intelligent Agent Technology can be found and applied to different aspects of our activities and working environment.

In the next four chapters, we will explore different AI applications areas which include: 5G and Intelligent Transportation, Smart City, Smart Campus, Smart Health, and Smart Education.

References

Alonso, E., et al. (Eds.). (2003). *Adaptive agents and multiagent systems: Adaptation and multi-agent learning.* Springer.

Baumann, J. (2000). *Mobile agents: Control algorithms.* Springer.

Descartes, R. (1998). *Meditations and other metaphysical writing* (D. M. Clarke, Trans.). Penguin Books.

Ferber, J. (1999), *Multi-agent systems: An introduction to distributed artificial intelligence.* Addison-Wesley.

FIPA. (2020). Retrieved May 10, 2020, from Official site: https://fipa.org/.

Furukawa, K., et al. (Eds.). (1999). *Intelligent agents: machine intelligence.* Oxford University Press.

Irwin, W. (2005). *More matrix and philosophy: Revolutions and reloaded Decoded* (Popular Culture and Philosophy). Open Court.

Jaworski, W. (2011). *Philosophy of mind: A comprehensive introduction.* Wiley-Blackwell.

Kolodziej, J., et al. (Eds). (2015). *Intelligent agents in data-intensive computing* (Studies in Big Data Book 14). Springer.

Lee, R. S. T. (2006). *Fuzzy-neuro approach to agent applications: From the AI perspective to modern ontology.* New York, Berlin: Springer.

Lee, R. S. T. (2019). COSMOS Trader—Chaotic neuro-oscillatory Multiagent financial prediction and trading system. *The Journal of Finance and Data Science, 5,* 61–82.

Lee, R. S. T. (2020). *Quantum finance—Intelligent forecast and program trading systems.* Springer.

Lee, R. S. T., & Loia, V. (2007). *Computational intelligence for agent-based systems.* New York, Berlin: Springer.

Murch, R., & Johnson, T. (1998). *Intelligent software agents.* Prentice Hall.

Nyvlt, M. J. (2011). *Aristotle and Plotinus on the intellect: Monism and dualism revisited.* Lexington Books.

Omicini, A., Zambonelli, F., Klusch, M., & Tolksdorf, R. (2001). *Coordination of Internet agents: Models, technologies, and applications.* Springer.

Plato. (1999). *Phaedo.* Gallop D. (Ed.). Oxford University Press.

Rothermel, K., & Popescu-Zeletin, R. (Eds.). (1997). *Mobile agents: Proceedings of the 1st International workshop (MA 1997).* Springer.

Russell, S., & Norvig, P. (2011). *Artificial intelligence: a modern approach* (3rd ed.). Pearson.

Shen W. (2001). *Multi-agent systems for concurrent intelligent design and manufacturing.* Taylor & Francis.

Tuchong. (2020a). *Intelligent agents and robotics.* Retrieved May 10, 2020, from https://stock.tuchong.com/image?imageId=473008785683513407.

Tuchong. (2020b). *Agents threats and challenges.* Retrieved May 10, 2020, from https://stock.tuchong.com/image?imageId=426647542907929196.

Tuchong. (2020c). *Shopping and bargaining agents.* Retrieved May 10, 2020, from https://stock.tuchong.com/image?imageId=260004796052013130.

9

Intelligent Transportation

Connected vehicle technologies are revolutionizing and democratizing transportation for safer, smarter, more responsible, and more accessible driving.
Dr. Dinesh Paliwal (CEO of Harman International, born 1957)

Abstract Transportation is an integral part of any society. It is closely related to lifestyle, range of activities and location, has goods and services available for consumption. Transportation advances have made it possible to change the way of life of social organizations, and therefore have a huge impact on civilization development. By integrating the latest 5G technology (fifth-generation cellular technology) with AI technology, the future of mobility is expected to change the way people transport goods because shared and autonomous vehicles can provide faster, cleaner, cheaper, and safer shipping options. This chapter begins with a general transportation system overview. Next, we study the major component of intelligent transportation—5G technology. Then, we study intelligent transportation system (ITS) and its potential applications followed by 5G-enabled ITS technology—V2X (vehicle-to-everything) with related technologies such as V2V (vehicle-to-vehicle) and V2I (vehicle-to-infrastructure) Technology. We also study 5G and AI technology integration for the implementation of new-age ITS applications that include smart cities, autonomous vehicles, intelligent traffic management systems, emergency services, and future-proof infrastructure.

R. S. T. Lee, *Artificial Intelligence in Daily Life*,
https://doi.org/10.1007/978-981-15-7695-9_9

History tells us that transportation has changed civilization. Transportation is an integral part of any society. It is closely related to lifestyle, range of activities and location, has goods and services available for consumption. Transportation advancements have made it possible to change the way of life of social organizations, and therefore have a huge impact on civilization development. By integrating the latest 5G technology (fifth-generation cellular technology) with AI technology, the future of mobility is expected to change the way people transport goods because shared and autonomous vehicles can provide faster, cleaner, cheaper, and safer shipping options.

We begin with a general overview of the transportation system and how it reshapes our society. Next, we examine the main component of intelligent transportation—5G technology and outline from 1 to 5G in the past half-century followed by 5G specifications analysis. Then, we will study *intelligent transportation systems (ITS)* and its potential applications, in addition to 5G-supported ITS technology-*V2X (vehicle-to-everything)* and related technologies, such as *V2V (vehicle-to-vehicle) and V2I (vehicle-to-infrastructure) technology.* After that, we study 5G and AI technologies integration in a new era of ITS applications, including smart cities, autonomous vehicles, intelligent traffic management systems, emergency services, and future-proof infrastructure, and how these applications reshape our daily life.

9.1   Transportation and Society

Transportation is an integral part of any society. It has been responsible for civilization development to satisfy transporting individuals and goods requirements since ancient times. This movement has changed the way we live and travel. In developed and developing countries, a large proportion of individuals travel to work, shop, and socialize every day. But transportation also consumes a lot of resources such as time, fuel, materials, and land (Comfort 2020).

The migration from rural habitats to urbanized habitats is different in developing countries. Many areas have been urbanized without obvious motorization and suburban formation. A small percentage of the population can afford cars, but cars greatly increase traffic congestion in intermodal transportation systems. They also generate a lot of air pollution, which poses major security risks and exacerbates inequality in society. Multimodal transportation systems for walking, cycling, motorcycles, buses, and trains can support high population densities.

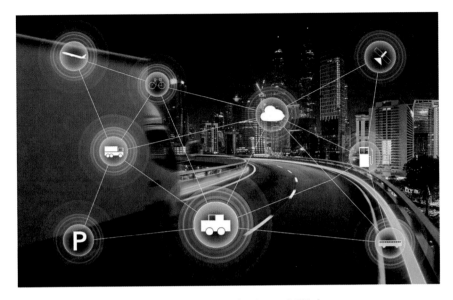

Fig. 9.1 Intelligent transportation systems (Tuchong 2020a)

Intelligent transportation system (ITS) (Bazzan and Klügl2013; Blokdyk 2020) is an advanced application designed to provide innovative services related to different transportation modes and traffic management so that users can be better informed, safer, coordinate, and use transportation networks more intelligently. ITS are used in technologies such as basic management systems (autonomous driving and car navigation), traffic-signal control systems, container management systems, variable message signs, automatic license plate recognition or high-speed cameras for monitoring safety CCTV systems, etc. to more advanced applications, such as traffic forecasting and management. Forecasting techniques are being developed to allow advanced modeling and comparison with historical baseline data. Figure 9.1 shows the ITS conceptual diagram with the integration of various supporting technologies such as cloud technology, big data, and AI.

9.2   From 1 to 5G Technology

Telecommunication, particularly the latest *5G Technology (fifth-generation cellular technology)* (Osseiran et al. 2016) with its low latency rate and promising 20 Gbps data rates provides a critical mass for ITS applications implementation.

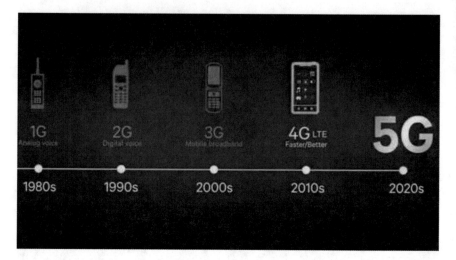

Fig. 9.2 Timeline from 1 to 5G technology

Before that, let us have a mobile communication technology history overview, from 1 and 5G and we will know why 5G is ITS critical mass. Figure 9.2 shows the timeline from 1 to 5G.

1G—The Rise of Mobile Revolution

The *first-generation mobile network (1G)* was created in 1979 by the Tokyo Telegraph and Telephone Corporation (NTT) in Japan. It had launched 1G to cover the whole territory of Japan by 1984.

The US government approved the first batch of 1G operations, and Motorola's DynaTAC became one of the first mobile phones to be widely used in the US in 1983. Other countries such as Canada and the United Kingdom also launched their own 1G networks a few years later.

However, 1G technology suffered a series of defects. The coverage was substandard and sound quality was inferior. There was no roaming support between various operators because different systems operate with different frequency ranges, there was no compatibility between systems. To make matters worse, the call was not encrypted, anyone using a radio scanner can answer the call. Despite these shortcomings and a high price of US$3,995 (about US$9,660 today), DynaTAC managed to attract a staggering 20 million global users by 1990. The success of 1G paved the way for the *second-generation mobile network also known as 2G.*

2G—The Era of GSM and SMS

The *second-generation mobile network (2G)* was launched in Finland in 1991 according to the GSM standard. This was the first time that calls could be encrypted, digital voice calls were clearer, static with reduced background noise. But 2G was not just telecommunications, it also laid the foundation for the cultural revolution. Users could send *text messages (SMS), picture messages,* and *multimedia messages (MMS)* on their phones for the first time. The analog past of 1G was replaced by the digital future presented by 2G that had led to large-scale adoption by consumers and businesses on an unprecedented scale.

Although the initial 2G transmission speed was only 9.6 kbit/s, operators are eager to invest in new infrastructure such as mobile cellular towers. By the end of the era, speeds can reach 40 kbit/s, and *EDGE* connections provide speeds up to 500 kbit/s. Despite its relatively slow speed, 2G had completely changed the business landscape and the world forever.

3G—The Revolution of Packet-Switching

The *third-generation mobile network (3G)* was launched by *NTT DoCoMo* (Japan) in 2001 and aims to standardize network protocols used by suppliers. This means that users could access data from anywhere in the world by standardizing *data packets* that drive network connections. This made international roaming services a reality for the first time.

The enhanced data transmission capabilities of 3G (four times faster than 2G) had also led to new services such as video conferencing, video streaming, and voice over IP (such as Skype). The *BlackBerry* was launched in 2002 and many of its powerful functions were realized through 3G connections. The 3G twilight era witnessed iPhone release in 2007, which meant that its network functions would be expanded as before.

4G—The Era of Data and Video Streaming

The *fourth-generation mobile network (4G)* as a *long-term evolution (LTE)* 4G standard was first deployed in Stockholm, Sweden, and Oslo in 2009. Subsequently, it was launched globally and made high-quality video streaming a reality for millions of consumers. 4G provides fast mobile Web access, i.e. up to 1 Gb per second for fixed users to promote gaming services, high-definition video, and HQ video conferencing.

The problem is that although the transition from 2 to 3G was as simple as switching SIM cards, mobile devices needed to be specifically designed to support 4G. By launching new mobile phones that support 4G, this had helped equipment manufacturers expand profits significantly. This was one reason why Apple became the world's first company with a market value of trillions of dollars. Although 4G is the current global standard, some areas are still plagued by network patching issues, and 4G LTE penetration is low. According to mobile data platform Ogury, for example, British residents can only access 4G networks 53% of the time.

5G—IoT and AI

Given the low 4G coverage in some regions, why has the focus shifted to 5G? In fact, 5G has been around for many years. In an interview with the Tech Republic, technology pioneer Mr. Kevin Ashton mentioned in his speech how to create the term *internet of things (IoT)* to convince Procter & Gamble (P&G) to use *RFID tag* technology in the 1990s. IoT is all the rage and will soon be touted as the next major digital revolution. It will witness billions of connected devices sharing data seamlessly across the globe. According to his description, a mobile phone is not only a phone, it's the IoT in our pocket; a number of network-connected sensors that help us accomplish everything from navigation, photography to communication, and more. IoT will see data moving out of the server center and into so-called *Edge devices,* such as Wi-Fi-enabled devices such as refrigerators, washing machines, and cars we are using nowadays.

By the early 2000s, developers knew that 3G or even 4G networks would not be able to support such networks. Since 4G's delay between 40 and 60 ms was too slow for real-time response, some researchers have begun to develop next-generation mobile networks.

The *National Aeronautics and Space Administration (NASA)* helped establish the *machine-to-machine intelligence (M2Mi)* corporation in 2008 to develop IoT and M2M technologies and 5G technologies needed to support this technology as well. In the same year, South Korea formulated a 5G R&D plan, and New York University created NYU wireless network focused on 5G in 2012.

The superior connectivity provided by 5G is expected to transform everything from banking to health care. 5G offers innovative possibilities such as remote emergency, remote medical, and even remote vital sign monitoring which can save lives.

South Korea's three major telecommunications operators—*KT, LG Uplus, and SK Telecom* to launch real-time commercial 5G services in December 2018 and promised to launch 5G nationwide in March 2019 simultaneously.

9.3 5G Specification

The *fifth-generation wireless technology* (5G in short) is an evolution in mobile technology from existing 4G LTE networks. It is designed specially to accomplish the current technological demands like huge data growth and approximately worldwide connectivity along with *IoT* increasing interest. Currently, it is the newest cellular technology in wireless technology which enlarges the wireless network speed greatly between other things. So, 5G-based wireless broadband connections data speed will be around 20 Gbps. This technology will also give additional bandwidth and superior antenna technology so that much more data will be transmitted over wireless systems.

The 5G technology is driven by the following specifications:

- Data rate is up to 10Gbps,
- Network energy usage reduction is 90%,
- Latency is 1 ms (millisecond),
- Maximum coverage,
- Availability is 99.999%,
- 100x devices can be connected for each area, and
- Bandwidth for each unit area is 1000x (Fig. 9.3).

Fig. 9.3 5G and IoT (Tuchong 2020b)

As we have seen, 5G poised to act as a future mobile network helping to make IoT a reality. This would not have been possible without the steady march of technological progress from 1G to the present day. As we can see why 5G is ITS critical mass of the intelligent city. Say, for example, it is almost impossible to implement a truly real-time base auto-driving system with a low latency of wireless data transfer and communication, let alone for the integration with other AI and data mining capability to ITS and intelligent city systems.

9.4   Intelligent Transportation System (ITS)

What is Intelligent Transportation System?

The concept of a smart city transforms the city into a digital society making life easier for citizens in all aspects, so the *intelligent transportation system (ITS)* has become an indispensable part of all. *Mobility* is a key issue in any city. Travel to schools, colleges, and offices or any other purpose using the transportation system to travel around the city. Citizens with ITS can save their time and even make the city smarter. ITS aims to improve traffic efficiency by minimizing traffic problems. It provides users with a priori information about traffic, local convenience real-time information, seats availability, etc. thereby reduce travel time in a safe and comfortable manner.

Today, ITS applications are widely accepted and used in many countries. This function is not limited to traffic congestion control and information, but also includes road safety and effective infrastructure use. Because of its unlimited possibilities, ITS has now become a joint field of work across disciplines, so many organizations worldwide have developed ITS applications to meet demand. For example, in Glasgow, ITS regularly provides daily commuters with regular information about public buses, time, seats availability, current and next location bus routes, time required to reach specific destinations, and passenger density in buses.

ITS Applications

ITS has the potential to relieve some of the most difficult problems affecting road transport today. Its applications, in general, have the capability to (Mouftah et al. 2018).

- Improve traffic flow to reduce congestion and minimize travel delays.
- Detect incidents quickly and respond to them appropriately.

- Improve safety to provide warning in advance before potential crash situations.
- Reduce pollution levels to improve air quality.
- Minimize environmental, highway, human impact factors contribute to accidents.

ITS can also make travel more convenient by providing commuters with accurate, timely information about traffic conditions on the network with available transport options. It can foster economic growth, improve mobility, enhance reliable travel time, and reduce energy consumption.

9.5 V2X Technology

9.5.1 5G-Enabled ITS Technology

As 5G technology becomes reality we will soon be seeing vast improvements in ITS capabilities. This will impact our day-to-day travel in 3 key areas (Miucic 2018; Mahmood 2020):

(1) *Vehicle-to-Vehicle (V2V) communications*

V2V (vehicle-to-vehicle technology) (Jurgen 2012; Miucic 2018; Mahmood 2020) is a communication system between road vehicles that organizes the driving environment. Over time, this smart driving solution and the smart city under construction will reduce traffic congestion and improve road safety. The vehicle is equipped with many sensors that can absorb all information about itself. Information includes vehicle speed, braking, driving direction, even steering wheel angle, road, traffic conditions, etc. where the driver is located; organized and sent to public spaces shared with other drivers. 5G will enable direct communication between vehicles without going through any network. They will share the traffic information, such as traffic jams, accidents, changes in road conditions, emergency vehicles traffic, adverse weather, etc. Autonomous vehicles will approach each other in groups of vehicles traveling in the same row or in the same direction. The obvious effect of instant messaging will redefine traffic optimization to improve road safety for all users. Figure 9.4 illustrates the V2V technology scenario.

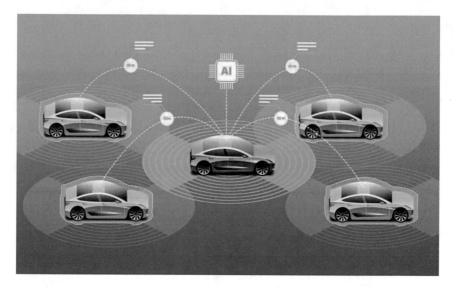

Fig. 9.4   Vehicle-to-Vehicle Technology (V2V) technology

(2) *Vehicle-to-Infrastructure (V2I) communications*

5G will enable communication between vehicles and their surrounding infrastructure such as traffic lights and cameras, drones, and freight sensors. The wireless technology used by existing systems is medium because communication is decentralized due to transmission through multiple networks and towers. Not only can 5G merge communications with lower latency, but it can also introduce more advanced and smarter applications. It will be more adaptable than anything we have seen so far (Jurgen 2012; Miucic 2018; Mahmood 2020). In ITS, *vehicle-to-infrastructure (V2I)* sensors can capture infrastructure data and provide users with real-time information such as traffic jams, accidents, road conditions, building areas, and parking facilities. Similarly, traffic management and monitoring systems can use infrastructure and vehicle data to set variable speed limits and adjust *traffic-signal phase and timing* (SPaT) to redefine fuel consumption and traffic optimization. The hardware, software, and firmware that facilitate communication between vehicles and road infrastructure are important components of all driverless car initiatives. Figure 9.5 shows V2I, V2V, cloud, and 5G technologies integration in ITS.

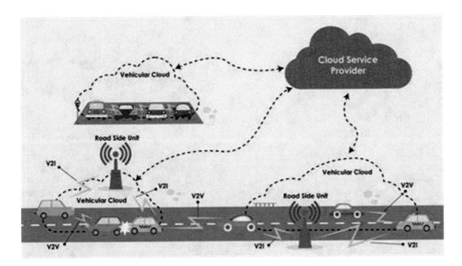

Fig. 9.5 V2I and V2V technology integration

(3) *Multimodal transportation*

5G will provide near-seamless communication when travelers use several transportation modes for a single journey. Let us say we drive an autonomous car to the bus station, take a bus to the city center, and hop onto a rental electric bike. Public transports are by digital payment (which can be connected securely within these separate systems) (Szyliowicz et al. 2016) that enable our commutes efficient and environmentally friendly.

9.5.2 Vehicle-To-Everything (V2X) Technology

V2X (or vehicle-to-everything) is *V2V (vehicle-to-vehicle), V2I (vehicle-to-infrastructure),* and *I2V (vehicle-to-infrastructure)* communication systems collaboration. The idea of this concept is to provide the driver with more information about the surrounding environment at any time while driving. Figure 9.6 shows the V2X technology scenario in a smart city (Mueck and Karls 2018; Miucic 2018; Mahmood 2020). V2X usually means that the vehicle is integrated with AI and all information can be obtained through sensors and the vehicle itself (driver's technical data and patterns), such as road infrastructure, traffic lights, route markings, etc. (referred to as V2I).

By using V2X technology in ITS, the information is recorded, organized, and sent to public spaces, as well as statistical information about weather and road conditions obtained from the Internet. In this public (or unified)

Fig. 9.6 Vehicle-to-Everything (V2X) Technology

space (or ecosystem), data analysis is drawn by AI, and it is drawn together with other vehicles. data (V2V). Therefore, a digital version of the driver and surrounding environment information will be created so that AI can track the driver's movement, nearby drivers and pedestrians, so that when immediate action is required, i.e. applying the brake pedal.

This is an uninterrupted process in which objects and activities are scanned along the route and updated in real time. Another benefit of the V2X system is that it is equipped with a better GPS route (to avoid traffic and bad weather), the nearest parking space (famous smart city function), and blind-spot perception function.

9.6   Potential Applications of ITS with 5G and AI Technology

ITS technologies are set to refine transportation systems and infrastructure that include traffic management, idle vehicle time, and relief congestion. The inclusion of modern 5G wireless networks and AI technology could enhance their operation and performance in several ways (Bazzan and Klügl 2013; Blokdyk 2020; Mouftah et al. 2018).

9.6.1 Smart Cities

One of the emerging concepts to upgrade wireless communications, advance automation, and IoT of smart cities. These smart cities would be able to collect and share data gathered from potentially millions of connected sensors built in city's infrastructure including intelligent transportation systems (Chow 2018). These would begin to further lead us toward truly intelligent cities with ultra-low latency networks for real-time data sharing between vehicles and transport infrastructure. As AI and automation technologies mature, it is likely they would rely heavily on 5G wireless networks to perform optimally.

9.6.2 Autonomous Vehicles (Auto-Driving)

The idea of autonomous vehicles (McGrath 2019) has been around far longer than any of the vehicles being developed. We are now in an age where driverless vehicles have become a reality.

These autonomous vehicles require data sharing almost real time for safety and effectiveness. They would likely become more popular once 5G networks begin to roll out in a not too distant future. They would be able to detect other vehicles, transport infrastructure, and pedestrians within sufficient time response to potential threats encountered. 5G networks have been a proposed solution to this problem. Figure 9.7 shows a scenario of an auto-driving system with AI and 5G technology integration. The latest AI-based low latency active vision would provide accurate real-time object recognition in high bandwidth, accurate location positioning feedback positions, and other related data.

9.6.3 Intelligent Traffic Management Systems

Vehicles data are required for the traffic transport system to function efficiently. This could mean to collect and process data from a potential thousands of connected vehicles on busy highways and from congested cities. ITS traffic management systems could benefit from 5G wireless networks increased capacity (Gordon 2015) to cope with the high demand. Reliability is another key factor in traffic management systems. Vehicles sit idle in front of traffic lights are no good if there is no connection between vehicles and traffic lights. Using 5G wireless networks would ensure that vehicles

Fig. 9.7 Auto-driving system with AI and 5G Technology (Tuchong 2020c)

and traffic management systems are in unceasing connection for safety and efficiency.

9.6.4   Emergency Response Services

When data mining is integrated AI with fast and low latency of 5G technology, it would provide a definitive solution for emergency services (Elhoseny and Hassanien 2019). These include empowering emergency service professionals to consolidate new technologies and services, e.g. ambulances with high-definition video communication, computed tomography (CT), X-ray scans to connect between vehicles and emergency departments. Network slicing is another way in which emergency services could use 5G wireless networks to their advantages. Emergency service vehicles using 5G wireless communications systems would be able to alert traffic management systems to modify traffic and provide optimal routes to destinations. Emergency service prioritization in smaller scales exist today and extend prospects with 5G.

9.6.5 Future-Proof Infrastructure

AI network infrastructure with 5G will take lesser time for system upgrade in the future. It used to be a long and arduous process but thanks to a combination of connectivity enhancement, shared standards, and an IoT focus wireless communication, this will no longer be the case (Sciarretta and Vahidi 2019). The smoother upgrade paths dispensed by future technologies provide stronger, up-to-date services such as security to protect both intelligent transportation systems and wireless networks. They would prevent hackers to exploit vulnerabilities as technologies and equipment are more advanced.

9.7 Case Study: Smart Car for ITS

We have studied various ITS applications in this chapter. ITS is having a significant effect in transportation applications such as auto-driving, emergency services, electronic toll collection, ramp meters, traffic light cameras, traffic-signal coordination, transit signal priority, and traveler-information systems.

Suppose you are a group member to design a next-stage *smart car* with AI, 5G, and ITS integration. Based on AI technology you have learnt in previous chapters that include.

- Computer vision (CV),
- Data mining (DM),
- Natural language processing (NLP),
- Ontological-based search engine (OSE), and
- Machine learning (ML).

Propose core functions and features of your designed *smart car* and explain how AI technology can be integrated with 5G technology for your dream *smart car* implementation (Fig. 9.8).

9.8 Conclusions

We discuss intelligent transportation system (ITS), one of the foremost AI and mobile communication technology application that affect our daily activities. ITS is the application of sensing, analysis, control, and communications technologies for ground transportation to improve safety, mobility,

Fig. 9.8 Smart car for ITS (Tuchong 2020d)

and efficiency. It includes a wide range of applications that process and share information to improve traffic management, ease congestion, minimize environmental impact, and increase transportation benefits to commercial users and the public. The success of ITS is highly dependent on computing advancement, networking, and mobile communication systems, particularly on the latest 5G technology to provide a seamless positioning, low latency, and high bandwidth network to communicate for rapid, accurate access critical information and data.

Toronto, Canada has recently embarked on an aggressive vision for a planned Quayside smart community along its waterfront. The overall project is an ambitious attempt to create a data-driven city powered by several technologies in which transportation is a key feature of the project. Private vehicles will be banned within the initial 12 acres district. All on-road vehicles will be autonomous, and goods will be delivered via freight robots in underground tunnels. IoT sensors installed at sidewalks to monitor pedestrian traffic and ensure real-time safety for road crossings. Cameras use AI to guide traffic flow. All relevant organized information communicated with pedestrians through an app. The extreme robust networks and IT services will be critical for this system to work. This smart city deployment presents a vision of the future where is completely dependent on ITS (Fig. 9.9).

Fig. 9.9 Future auto-driving with 5G + AI technology (Tuchong 2020d)

References

Bazzan, A. L. C., & Klügl, F. (2013). *Introduction to intelligent systems in traffic and transportation* (Synthesis Lectures on Artificial Intelligence and Machine Le). Morgan & Claypool Publishers.

Blokdyk, G. (2020). *Intelligent transportation system a complete guide—2020 edition* (Kindle edition). 5STARCooks.

Chow, J. (2018). *Informed urban transport systems: Classic and emerging mobility methods toward smart cities.* Elsevier.

Comfort, P. (2020). *The future of public transportation* (Kindle edition). Comfort Consulting World Wide.

Elhoseny, M., & Hassanien, A. E. (2019). *Emerging technologies for connected Internet of vehicles and intelligent transportation system networks: Emerging technologies for connected and smart* (Systems, Decision and Control Book 242). Springer.

Gordon, R. (2015). *Intelligent transportation systems: Functional design for effective traffic management* (2nd ed.). Springer.

Jurgen, R. K. (2012). *V2V/V2I Communications for improved road safety and efficiency.* SAE International.

Mahmood, Z. (2020). *Connected vehicles in the Internet of Things: Concepts, technologies and frameworks for the IoV.* Springer.

McGrath, M. E. (2019). *Autonomous vehicles: Opportunities, strategies and disruptions: Updated and expanded second edition* (Kindle edition). Amzaon.com.

Miucic, R. (2018). *Connected vehicles: Intelligent transportation systems* (Wireless Networks). Springer.

Mouftah, H. T., et al. (2018). *Transportation and power grid in smart cities: Communication networks and services*. Wiley.

Mueck, M., & Karls, I. (2018). *Networking vehicles to everything: Evolving automotive solutions*. DelG Press.

Osseiran, A., et al. (2016). *5G mobile and wireless communications technology*. Cambridge University Press.

Sciarretta, A., & Vahidi, A. (2019). *Energy-efficient driving of road vehicles: Toward cooperative, connected, and automated mobility* (Lecture Notes in Intelligent Transportation and Infrastructure). Springer.

Szyliowicz, J. S., et al. (Eds). (2016). *Multimodal transport security: Frameworks and policy applications in freight and passenger transport* (Comparative Perspectives on Transportation Security series). Edward Elgar Publishing.

Tuchong. (2020a). *Intelligent transportation systems*. Retrieved May 10, 2020, from https://stock.tuchong.com/image?imageId=475108835712696440.

Tuchong. (2020b). *5G and IoT*. Retrieved https://stock.tuchong.com/image?imageId=918242515924549693. Accessed 10 May 2020.

Tuchong. (2020c). *Auto-driving system with AI and 5G Technology*. Retrieved May 10, 2020, from https://stock.tuchong.com/image?imageId=259828272896147689.

Tuchong. (2020d). *Smart car for ITS*. Retrieved May 10, May 2020, from https://stock.tuchong.com/image?imageId=257312470032711887.

10

Smart Health

I have seen that technology has contributed to improved communication, that it's contributed to better health care, that it's contributed to better food supplies, that it has contributed to all the basic human needs.
Prof. John Warnock (Co-founder of Adobe System, born 1940)

Abstract New technologies such as AI, big data, and the latest 5G technology have many impacts on our daily activities. Today's healthcare systems have recognized the advantages of smart technology that can improve healthcare quality, thereby turning traditional technologies into smart health care. Smart medical technology can define better monitors, diagnostic tools for patient treatment, and devices to improve the quality of life. Key smart health concepts include eHealth and mHealth services, electronic record management, smart home medical services, and connected smart medical devices using the latest IoT (internet of things) and wearable computing technologies. In this chapter, we start with the impact and challenges of medical technology in the twenty-first century. Next, we outline IoT and wearable computing technologies, then analyze how such technologies reshape the blueprint for smart health. We also explore the latest wearable medical devices inventions from smartwatches to wearable biosensor technology. After that, we study two most important AI technologies used in health care: (1) Health chatbot; (2) Robot-Assisted Surgery (RAS) technology.

R. S. T. Lee, *Artificial Intelligence in Daily Life*,
https://doi.org/10.1007/978-981-15-7695-9_10

New technologies such as AI, big data, and the latest 5G technology have many impacts on our daily life. Today's healthcare systems have recognized the advantages of *smart technology* that can improve healthcare quality, thereby turning traditional technologies into *smart health care*. Smart medical technology is defined by technologies that can define better monitors, diagnostic tools for patient treatment, and devices that can improve the quality of life. The key smart health concepts include eHealth and mHealth services, electronic record management, smart home medical services, and connected smart medical devices using the latest *IoT (internet of things)* and *wearable computing technologies.*

In this chapter, we start with the impact and challenges of medical technology in the twenty-first century. Next, we outline IoT and wearable computing technologies, then analyze how such technologies reshape the blueprint for smart health. We also explore the latest wearable medical devices inventions from smartwatches to monitoring heart rate, blood pressure to wearable biosensor technology to monitor our health condition and athletic performance. After that, we will study the two most important AI technologies used in health care: (1) health chatbot and AI doctor; (2) *Robot-assisted surgery (RAS)* technology.

10.1   Healthcare in the Twenty-First Century

Global healthcare system faces enormous challenges in the twenty-first century. According to *Frost & Sullivan's Social Innovation Report on Healthcare* (Frost & Sullivan 2016), the population aged 65 years or over is expected to grow from 524 million in 2010 to nearly 1.5 billion by 2050, while OECD National long-term care will nearly double GDP of 14%. Although the base in BRIC countries is relatively low, about 2.5%, it is expected to grow by about 10%. With advances in technologies such as the internet of things (IoT), wearable technology, artificial intelligence, and data mining, the future of health care is changing globally (Reis 2016; Hassanien et al. 2018).

Traditional health care was focused on medicines and devices in the past half-century. It has turned to integrated services and added value around products or equipment in the past 5 years. Emerging technologies will continue to develop digital and AI experience in 2020 and beyond. Today's patients are knowledgeable and often use popular search engines as the first reference site. In addition to obtaining medical knowledge, they also obtained a large amount of reference materials to consult physicians and guide further research. The reality is that there are different kinds of patients,

Drugs and Devices 1980 - 2010	Services 2010- 2016	Digital Experience 2016 -
	Diagnostics, Health Management.	
Focus on the DISEASE	Prevention. Home Monitoring, Wearables	Data Driven Ownership and Empowerment, Wellness, Prediction Personalization Virtualization

Fig. 10.1 Healthcare focus changes in the past 40 years (Frost & Sullivan 2016)

who entice healthcare companies to personally understand their interactions with products and actively manage their therapies and health digitally. The combination of new technologies and knowledge has strengthened the doctor–patient relationship and provided innovations in healthcare products and services (Fig. 10.1).

10.2 Internet of Things (IoT)

What is IoT?

The internet of things (IoT) refers to a network of physical objects capable of collecting and sharing electronic information (Reis 2016). It includes various smart devices from machines that transfer data from production processes to industrial machines, to sensors that track information about the human body. Objects in IoT (which can be humans or smart devices) can be individuals implanted with heart monitors, farm animals with biochip transponders, cars with sensors built in to warn drivers, or any other natural or human sensors. Assign objects to Internet Protocol (IP) addresses to transfer data over the network. Organizations from all walks of life are increasingly using IoT to improve operational efficiency, better understand customer needs, provide value-added services, and improve decision-making and business value.

The IoT ecosystem consists of web-based smart devices that use embedded systems such as processors, sensors, and communication hardware to collect, send, and process data obtained from the environment. IoT devices share sensor data they collect by connecting to IoT gateways or other edge devices. In these devices, data is sent to the cloud for analysis or analyzed locally.

Sometimes, these devices communicate with other related devices and actions based on the information obtained from each other. Although individuals can interact with devices, e.g. set up, provide instructions, or access data, these devices perform most of the work without human intervention. Figure 10.2 shows a conceptual diagram of IoT in a smart city used to connect information and knowledge from different devices.

Impacts of IoT to Our Daily Life

IoT can help us live and work smarter. In addition to providing smart devices to automate homes, IoT is also critical for businesses. It provides the company with a real-time view of how the system operates and provides insights from mechanical performance to supply chain and logistics operations. It enables companies to automate processes and reduce labor costs. It reduces waste, improves delivery services, and increases the transparency of customer transactions.

Generally speaking, IoT technology uses sensors and other devices in the agricultural, manufacturing, transportation, and utility industries to enable digital transformation. Home automation companies can use IoT to monitor

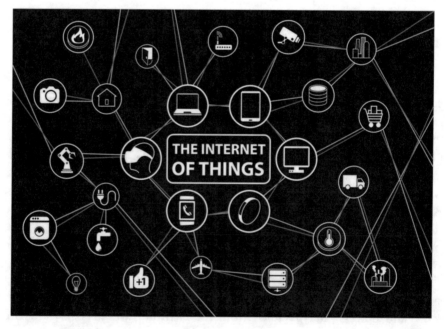

Fig. 10.2 Internet of things (IoT) (Tuchong 2020a)

and manipulate mechanical and electrical systems in buildings. The ability of IoT to monitor the operation of surrounding infrastructure is also a factor. Therefore, IoT has become one of the most foremost technologies in our daily activities, and as more and more industries realize the potential of IoT devices to remain competitive, IoT will continue to do so.

10.3 Wearable Technology

Wearable technology (WT) (Tong 2018), also known as a *wearable device*, is an electronic device that can be worn as an accessory, embedded in clothes, implanted in the user's body, or even tattooed onto our skin. They are hands-free devices driven by a microprocessor (sometimes with biosensors) and have the function of sending and receiving data via the Internet for actual use.

In fact, wearable technology is not a new idea in human history. It has emerged since spectacles development in the thirteenth century. From about AD1500, humans began to wear watches. Modern wearable technology is defined as connecting the microprocessor to the Internet. The growth of mobile networks has driven wearable technology development. Fitness activity trackers are the first wave of wearable technology to catch up with consumers. Then, the watch becomes a mobile computer with powerful mobile applications. Bluetooth headsets, smartwatches, and web-enabled glasses enable users to receive data from Wi-Fi networks. The gaming industry has added virtual and augmented reality headsets.

The rapid adoption of such devices puts wearable technology at the forefront of IoT. The focus of current wearable technology development seems to have shifted from consumer accessories to more professional and practical applications. Microchip implants are used to replace keys and passwords. The chip embedded in the fingertip uses near-field communication (NFC) or radio-frequency identification (RFID), alike the chip used to look for pets (Fig. 10.3).

10.4 Healthcare Wearable Technology

10.4.1 Wearable Technology and Health

Wearable technology products have been widely used in medical and health-care applications in the past few years. The technology is frequently used to monitor users' health. Given that the device is in close contact with the

Fig. 10.3 Wearables in VR headsets (Tuchong 2020b)

user, it can collect data easily. The first wireless electrocardiogram (ECG) was invented in 1980 and showed a rapid growth trend in research based on textiles, tattoos, patches, and contact lenses (Tong 2018; Dey et al. 2019).

Wearables can be used to collect user's health data include.

- Blood pressure,
- Calories burned,
- Exercise time and frequency,
- Heart rate,
- Physical strain,
- Release of certain biochemicals, and
- Seizures.

10.4.2 Healthcare Sensing Technology Applied to Wearable Devices

The latest developments in medical equipment indicate that surveillance is the main trend in homes and not specific locations in hospitals. According to this trend, medical technology is developing from the field of doctors and specialists to the general public, from hospitals to homes and mobile environments. Wearable devices and related services will play an increasingly important role. For example, small and inexpensive wearable devices can now be used to measure biomedical measurements easily such as electrocardiograms (ECG) and blood oxygen saturation that are measured using the device in hospitals. As a result, new medical services using measurement data can

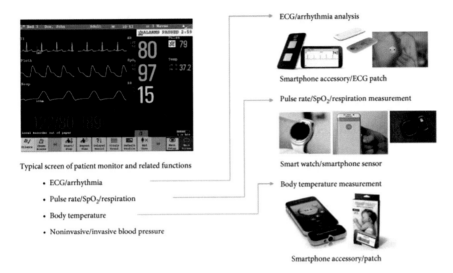

ECG/arrhythmia analysis

Smartphone accessory/ECG patch

Pulse rate/SpO₂/respiration measurement

Smart watch/smartphone sensor

Typical screen of patient monitor and related functions

Body temperature measurement

- ECG/arrhythmia
- Pulse rate/SpO₂/respiration
- Body temperature
- Noninvasive/invasive blood pressure

Smartphone accessory/patch

Fig. 10.4 Patient monitoring system for wearable devices (Cho 2019) (CC-pdm-1.0)

also be introduced. Figure 10.4 (Cho 2019) shows a typical screen of the patient monitoring system used by the hospital and lists its measurement functions. As shown, we can now use a small handheld device or a wearable device to record and analyze ECG signals, and measure heart rate, SpO2, respiratory rate, and body temperature.

10.4.3 Example of Wearables for Health Care

ADAMM Wearable Technology

ADAMM is a wearable technology of Health Care Originals (Healthcare-orginals 2020), which helps to understand and monitor asthma more easily. It consists of three parts: a wearable device that can detect precursor asthma symptoms, a smartphone application that can manage the basics of asthma control, and a portal that can detect treatment effects. There are about 300 million asthma patients worldwide benefited from this wearable device. The device and application will be able to alert users when an asthma attack, diary, treatment plan, display, and track symptom treatment information. Health Care Originals is a leading digital health startup that integrates IoT technology to enhance users' understanding and management of health and medical conditions.

Wearable Sweat Sensor Informs Athletes of Water and Electrolyte Loss

A team of researchers recently developed a waterproof, bandage-like sweat sensor to advise users when to replenish electrolytes and fluids. This innovative patch collects and analyzes athletes' perspiration in any environment and even swimming (Reeder et al. 2019). There are tiny holes on the bottom surface of the patch, which can penetrate the device as described in *Science Progress*. Each of these holes contains its own permeation analysis technique, as well as various indicators used to analyze when the wearer needs hydration or electrolytes. Sweat contains salt, sugar, hormones, and other valuable information, but few studies have used it as a measure of physical condition. The patch has a diameter of 1.5 in. and can stick to the user's skin. Device sensors analyze the information and portray it to the user's smartphone report. The device is also capable of wireless charging, making it very user friendly. The micropore in the patch center collects the sweat sample and fills the channel wound on the device surface. The collected samples are mixed and interact with the chemicals in these microchannels to change the color of the fluid (Fig. 10.5).

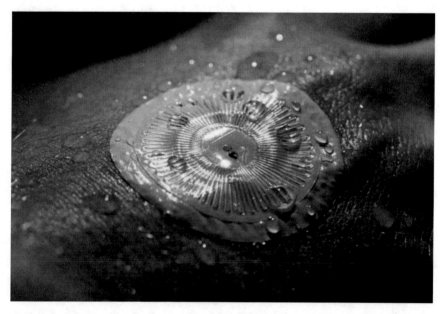

Fig. 10.5 Wearable sweat sensor (Reproduced with permission by Dr. Philipp Gutruf of The University of Arizona)

Current Health AI Wearable Device

Current Health's (Currenthealth 2020) AI wearable device, which can measure multiple vital signs has recently received FDA approval to allow patients to use at home. The Edinburgh, a Scotland-based company obtained permission to use AI-enabled devices for monitoring patients in hospitals in February 2019, recent approvals mean it can now also be used at home during doctor visits.

The wireless device *Current* can measure the patient's pulse, respiration, oxygen saturation, body temperature, and activity. *Current* provides doctors with real-time updates on patients' health, enabling them to quickly manage complications. The technology uses machine learning to analyze the collected data to detect problematic data changes. Patients with chronic obstructive pulmonary disease (COPD) and heart disease most often use *Current* because of the high hospitalization rate with these two diseases and the leading cause of death in the United States. The user can see the process as a serpentine channel with color indication. The length of the channel turns blue, representing the level of loss per breath (indicating a reduction in hydration). The other chamber also went through a similar process but changed from light pink to dark red indicating electrolyte loss.

Wearable Peritoneal Dialysis Device

AWAK Technologies (AWAK 2020) is a Singapore-based medical technology company that received FDA breakthrough device designation for its wearable and portable dialysis devices recently. AWAK peritoneal dialysis equipment or AWAK PD uses AWAK's patented adsorbent technology to provide a convenient dialysis method for patients with kidney disease.

Peritoneal dialysis (PD) is a common method of treating renal failure while hemodialysis is another method. Hemodialysis requires patients to visit dialysis centers for treatment by professional healthcare personnel, which causes inconvenience to patients. Patients who choose to use peritoneal dialysis can have more control according to situations, but this method still requires supervision.

The new AWAK device is designed to make the peritoneal dialysis process more convenient. The AWAK PD tool enables patients to perform dialysis, eliminating inconvenience caused by treatment time and going to/from the dialysis center. If there are no approved alternatives on the market currently, or if the technology has a huge advantage over existing alternatives, FDA considers the technology to be a breakthrough device. These names were created to speed up the process when these beneficial products are placed on the market.

After reviewing the safety tests conducted by patients using the device, the FDA believes that the AWAK PD device should have this status. The trial was conducted at Singapore General Hospital, the largest acute tertiary hospital in Singapore. This preliminary study involved 15 adults aged 21–80, who received conventional peritoneal therapy. Each patient competed with AWAK PD for nine dialysis treatments over a 3-day course, each course lasts for 3.5 h. The results of this study showed that AWAK PD can remove accumulated waste and fully filter blood, and patients did not experience adverse events during treatment.

AI and Wearable Sensor to Detect Heart Disease

According to preliminary data from a new study published in the American Academy of Cardiology's Annual Science in 2019, the Apple Watch app was able to detect atrial fibrillation (AFib) among a small group of users and remind them of irregular heart rhythm meetings in New Orleans (Dey et al. 2019; Tong 2018) (Fig. 10.6).

Fig. 10.6 Wearable iWatch with biosensor (Tuchong 2020c)

The prospective researcher is a single-arm study at Stanford University School of Medicine, which aims to provide funding to those with irregular pulse monitoring capabilities and then diagnoses AFib by electrocardiogram (ECG). The app generates a tachometer (time point between hot jumps) by using Apple Watch's existing light sensor. The algorithm analyzes whether the generated tachometer has irregularities. If an irregular rhythm is detected, a notification sent to the watch is triggered.

The notification recommends that users contact the research doctor through this application. The entire study cohort consisted of 419,297 participants, of which 2,161 participants sent notifications via Apple Watch, and based on the clinical follow-up results after notification, 450 participants wore ECG patches (and returned for analysis). 2,161 patients received a notification to complete the 90-day survey indicating that approximately 21% of the study cohort had hypertension and approximately 5% had diabetes. According to the introduction, for irregular pulse notifications, the positive predictive values of the travel chart are 0.71 (CI. 97.5%, 0.69–0.74) and 0.84 (97.5% CI, 0.76–0.92). Regardless of whether it has been evaluated by a virtual doctor, about half of the participants who have received the notification continue to seek medical assistance. The study also reported that young participants (under 40 years old) had a lower notification rate (1.6%), while elder participants (over 65 years old) had a higher notification rate (3% or higher). The overall irregular pulse notification rate is low at 0.52% (0.49–0.54).

10.5 Health Chatbot

Digital personal assistants (DPA) or *chatbots* can help doctors, nurses, patients, or their families in countless situations. Better patient pathway organization, medication management, emergency, or emergency assistance provide solutions to simple medical problems. These are all possible conditions for chatbots to intervene and reduce medical professionals' workload. Health chatbots are also used to solve specific healthcare issues. Researchers can also successfully monitor patient satisfaction, cancelation, non-appearance, and completed exams by using the app. In some cases, health chatbots can also link patients to clinicians for diagnosis or treatment. The general idea is that a smart algorithm that speaks or texts may become the initial point of contact for primary care in the future without having to contact a doctor, nurse, or any medical professional for each of their health issues. If the health chat robot cannot provide a satisfactory answer, it will transfer the case to a real medical professional.

GYANT (GYANT.com Inc.; GYANT 2020) is a health chat robot that requires patients to understand their symptoms, and then sends data to the doctor in real time for diagnosis and prescription. In addition, they not only provide services for English-speaking patients, GYANT can also communicate with users in Spanish, Portuguese, and German. The company reported again that they prompted more than 785,000 people in Latin America to successfully complete pre-diabetes screening in March 2019. Since then, more than 174,000 at-risk individuals from low-income groups have participated in weekly glucose tests at local pharmacies.

10.6 Robot-Assisted Surgery (RAS) Technology

The aging population causes a shortage of doctors. Hospitals and health systems are investing in robotic systems for surgery and telemedicine to improve patient capabilities and geographic coverage to meet the growing demand for healthcare.

Robot-assisted surgery (aka RAS) (Jarc and Nisky 2015) is a widely used technique, and thousands of surgeons use the *da Vinci Surgical System* (Intuitive Surgical in Sunnyvale, California, Inc.) performs surgery more than 500,000 operations per year. The success of RAS and other emerging technologies is due to its ability to treat patients with low invasiveness safely and effectively, because surgeons and robotic interactive surgery which not only brings opportunities for neuroscientists to develop our understanding of sensorimotor behavior, but also improves RAS technology; surgical training paradigms, and ultimately allows patients to experience RAS treatment. Figure 10.7 shows the RAS schematic diagram.

During RAS, the surgeon sits on the console, views the surgical area in three dimensions, and uses the master manipulator to control the instruments in the patient. The components of RAS are shown in Fig. 10.7c. The surgeon's motor system generates commands to cause hand movements and interacts with the main manipulator as input to the remote operating system that controls the instrument, or an endoscope fixed by a mechanical arm. The instrument or endoscope then interacts with the environment (i.e. the patient's tissue). In each step, the surgeon perceives all behavior aspects, and the robotic system uses visual, tactile, and auditory information channels. These, in turn, are combined with the internal environment and effective copy representation to provide surgeons with status estimates

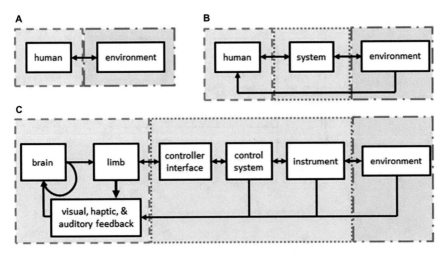

Fig. 10.7 Schematic diagram of robot-assisted surgery (Jarc and Nisky 2015)

to drive subsequent operations (including online corrections and new operations) and make strategic decisions about the next steps of surgical procedure. For example, to repair cancerous lymph nodes, surgeons use instruments to palpate carefully, dissect, and remove unhealthy tissues without destroying nearby structures.

RAS includes multiple parallel feedback loops around the surgeon's motor system as shown in Fig. 10.7c. Each of these parallel circuits is an opportunity to measure the surgeon's behavior or apply structural interference. Many experimental paradigms require these measures or disturbances such as cognitive reasoning, motor control, sensory processing, learning, and adaptation. During RAS, the surgeon's behavior can be measured in multiple stages, including hand movement, instrument movement, and surgical field of view. Similarly, by changing the motion or force applied to the main manipulator, the motion or visual feedback of the surgical instrument can disturb the surgeon. Three system categories are available for RAS research simulators, research platforms, and clinical systems. As mentioned above, all of these courses have achieved three key goals and can serve as a good experimental platform for human neuroscience research-task complexity, system fidelity, and user expertise as shown in Fig. 10.8 (Jarc and Nisky 2015).

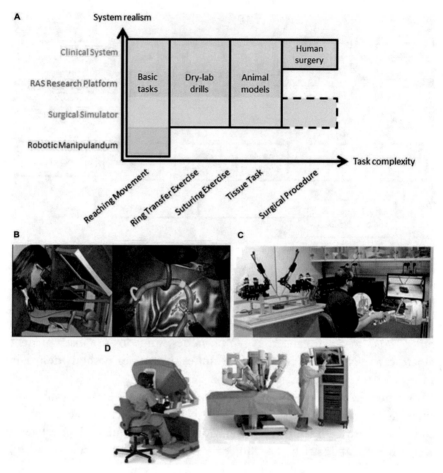

Fig. 10.8 Robot-assisted Surgery (Jarc and Nisky 2015) (CC BY 4.0)

10.7 Case Study: Can AI Replace a Real Doctor?

In this chapter, we discuss various AI and IoT-related medical systems and wearable devices to improve our health conditions and quality of life, ranging from wearable device to detect our heart conditions and asthma attack risk levels, to perform robotic surgical remote operations.

One might ask: Can AI and related technologies such as IoT, robotic, and data mining technology replace a physical doctor for patient diagnosis and/or surgery operations someday?

To answer this question, we need to explore what AI (e.g. robotic doctor) can and cannot be done at the current technology level, and then whether such areas can be achieved in the near future.

Fig. 10.9 Robot doctor (Tuchong 2020d)

Form a group of 4–5 students:

1. Explore the capability and requirements for a physical doctor include general medical doctors, medical specialists, and surgery doctors.
2. For each capability/requirement discuss and explain whether existing AI and related technology such as IoT, data mining, ontological search engine, etc. can be applied to replace a real doctor.
3. If not, can it be achieved in the near future? Why?
4. In addition to AI technology, what other aspects (e.g. ethics issue) we need to explore and consider when we try to implement such AI based robotic systems (Fig. 10.9).

10.8 Conclusion

We study one of the foremost AI applications—smart health and IoT (internet of things), a core technology that forms a critical mass of smart health technology. Healthcare system has recognized AI advantages and related technologies to improve healthcare quality, turning traditional into

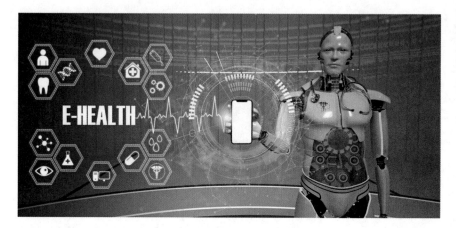

Fig. 10.10 AI and smart health (Tuchong 2020e)

smart health care. Smart Health is aimed to provide better monitoring, diagnostic tools, and patients treatment, and devices to improve the quality of life for everyone.

Can AI replace human medical doctor someday? Meantime it cannot. The main reason behind is that AI, IoT, and mobile technology today can help us a lot in smart health mainly focus to collect medical information, such as heartbeat rate, blood pressure, or other biometric information; powerful cloud and data mining systems with the knowledge to store and data-mine high-level knowledge for medical diagnosis. But it still (at least meantime) needs human experts (medical doctors) to consolidate information together with their medical knowledge and experiences to diagnosis.

Nevertheless, smart health technology especially various wearable medical devices such as heart disease or asthma monitoring device provide an excellent job for health status monitoring, personal and existing medical systems management are of benefit to the public (Fig. 10.10).

References

AWAK Technology. (2020). Retrieved May 14, 2020, from Official site: https://awak.com/.

Currenthealth. (2020). Retrieved May 14, 2020, from Official site: https://currenthealth.com.

Cho, J. (2019). Current status and prospects of health-related sensing technology in wearable devices. *Journal of Healthcare Engineering*, 2019, 3924508–8. https://www.ncbi.nlm.nih.gov/pmc/articles/PMC6604299/.

Dey, N., et al. (2019). *Wearable and Implantable Medical Devices: Applications and challenges*. Academic Press.

Frost & Sullivan. (2016). *Social innovation in healthcare whitepaper*. https://www.hitachi.eu/sites/default/files/fields/document/sib/whitepapers/social_innovation_in_healthcare_whitepaper1.pdf

GYANT. (2020). Retrieved May 14, 2020, from Official site: https://gyant.com/.

Hassanien, A. E., et al. (2018). *Medical Big Data and Internet of medical things: Advances, challenges and applications*. CRC Press.

Healthcareorginals. (2020). Retrieved May 10, 2020, from Official site: https://healthcareoriginals.com.

Jarc, A. M., & Nisky, I. (2015). Robot-assisted surgery: An emerging platform for human neuroscience research. *Frontiers in Human Neuroscience*, 9(June), 315. https://www.frontiersin.org/articles/10.3389/fnhum.2015.00315/full.

Reeder, J. T., et al. (2019). Waterproof, electronics-enabled, epidermal microfluidic devices for sweat collection, biomarker analysis, and thermography in aquatic settings. *Science Advances*, 5 Jan 2019: 5(1). https://doi.org/10.1126/sciadv.aau6356.

Reis, C. I. (2016). *Internet of Things and advanced application in healthcare*. Medical Information Science Reference.

Tong, R. (Ed.). (2018). *Wearable technology in medicine and health care*. Academic Press.

Tuchong. (2020a). *Internet-of-Things (IoT)*. Retrieved May 14, 2020, from https://stock.tuchong.com/image?imageId=260759670914023604.

Tuchong. (2020b). *Wearables in VR headsets*. Retrieved May 14, 2020, from https://stock.tuchong.com/image?imageId=473309502113710129.

Tuchong. (2020c). *Wearable watch with biosensor*. Retrieved May 14, 2020, from http://stock.tuchong.com/image?imageId=427991103167398062.

Tuchong. (2020d). *Robot doctor*. Retrieved May 14, 2020, from http://stock.tuchong.com/image?imageId=903578380886081574.

Tuchong. (2020e). *AI and smart health*. Retrieved May 14, 2020, from http://stock.tuchong.com/image?imageId=919946896381837450.

11

Smart Education

Nothing is more important, certainly during these times of artificial intelligence, than our public education. And as it continues to grow and evolve, I think you and I know this is going to be critical that we are constantly training and retraining and creating these next-generation jobs.
Mr. Marc Benioff (Internet entrepreneur, born 1964)

Abstract Education, especially public education using AI is more important than ever. This is because not only we need to reposition and upgrade our work skills, but more importantly, we should use this extraordinary technology to acquire knowledge, learn from others, and collaborate with all talents worldwide to acquire new knowledge. The richness of alliance is not only beneficial to us, but also to the overall mankind. This chapter begins with smart education progress in the past decades. Next, we examine the smart education model—a four-tier framework of smart pedagogies and key features of smart learning environment. Then, we study the two latest R&D AI-based smart education applications: AI language learning robots and VR-AR teacher to elaborate how AI technology we learnt in previous chapters such as machine learning, NLP technology, ontological knowledge base, VR, and AR can be integrated to provide a new age of smart education.

Mr. Marc Benioff's statement is correct because education, especially public education using AI is more important than ever. This is because not only do we need to reposition and upgrade our work skills, but more importantly, we should use this extraordinary technology to acquire knowledge, learn from others, and collaborate with all talents worldwide to acquire new sknowledge. The richness of alliance is not only beneficial to us, but also to the overall mankind.

In this chapter, we explore an important AI topic in daily activities—*smart education*. First, we review its progress for the past decades. Then, we analyze its model—a four-tier framework of smart pedagogies and key features of the smart learning environment. Next, we examine two latest R&D on AI-based smart education applications: AI language learning robots and VR-AR teacher to integrate AI technology we learnt from previous chapters such as machine learning (especially on reinforcement learning), NLP technology, ontological knowledge base, VR, and AR to provide a new age of smart education.

11.1   Smart Education—An Introduction

With the advancement of technology, anything can be instrumented, inter-connected, and integrated into intelligent design, as is education. In recent years, smart education has attracted widespread attention in projects carried out globally (e.g. Chan 2002; Choi and Lee 2012; Hua 2012; IBM 2012; Mäkelä et al. 2018; Zhu et al. 2016; Kobayashi et al. 2017).

Malaysia's *Smart School Implementation Plan,* launched the first and the *Smart Education Project* in 1997 (Chan 2002). In this project, smart schools were implemented to improve the education system to meet the challenges of the twenty-first century.

Singapore implemented its *Smart Country Master Plan* (2010), in 2006, in which *smart education* was supported by various latest technologies such as IoT (Internet of Things), AI, and 5G (Hua 2012). Eight *smart schools* focused on creating diverse learning environments were established in the plan.

The *Finnish Board of Education* also implemented a smart education project in 2011, with a continuous system learning solution (SysTech). The project aims to promote learning in the twenty-first century through user-driven motivational learning solutions (Mäkelä et al. 2018).

The Australian government worked with IBM to design and implement an intelligent, multidisciplinary intelligent education system (IBM 2012), to connect all schools, higher education institutions, and vocational training institutions in 2012.

At the same time, South Korea government have also launched their smart education project (Choi and Lee 2012) and the United Arab Emirates (UAE) government launched their smart learning program, the so-called *Mohammed Bin Rashid (MBRSLP),* which aims to establish a new learning and cultural environment in national schools.

The New York Department of Education in the US launched the *Smart School Program,* in 2014, which aims to integrate the latest IT technology into the classroom (New York Smart School Committee Report 2014).

In Africa, the *ICT Transformation Education Project* (2019), was launched in 2015, with the aim to promote human and social development of African countries through the use of information and communication technology (ICT) for education. It is part of the cooperation between UNESCO and the Korea Trust Fund. The first phase of the project was implemented in Mozambique, Rwanda, and Zimbabwe from 2016 to 2019. The second phase will take place in Côte d'Ivoire, Ghana, and Senegal from 2020 to 2023.

Overall, the focus on smart education and development has become a new trend in global education.

11.2 What is Smart Education?

There is no clear and unified definition of *intelligent education* and *intelligent learning. Intelligent education* is a term describing education and learning in the new AI era and the digital age which has attracted the attention of many researchers. The goal of *smart education* (and *smart learning*) is to train smart learners by adopting the latest IT, communication, and AI technologies to meet the work and life needs of the twenty-first century.

Multidisciplinary researchers and education professionals are constantly studying the concepts of intelligent education and learning. Hwang (2014) believed that intelligent learning is ubiquitous learning with context awareness. Figure 11.1 shows the intelligent learning environment proposed by Hwang (2014).

The proposed smart education framework by Hwang (2014), consists of seven major components:

1. *Learning status detecting modul*e—to detect learners' real-world status and environmental contexts.
2. *Learning performance evaluation module*—to evaluate and record learners' performance by conducting online tests or in real world.

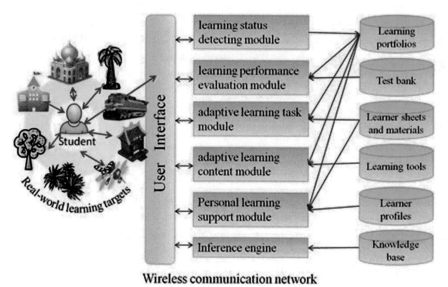

Fig. 11.1 Smart learning environment proposed by Hwang (2014) (CC BY 4.0)

3. *Adaptive learning task module*—to assign learning tasks to learners based on learning progress, learning performance, personal factors, and learning objectives in all disciplines.
4. *Adaptive learning content module*—to provide learning materials to learners.
5. *Personal learning support modul*e—to provide learning support to learners based on learning needs.
6. *Database module*—to keep learners' profiles, learning portfolios, learning sheets (i.e. the sheets that present the learning tasks for each subject unit or learning topic), learning materials, test items, and learning tools.
7. *Inference engine and a knowledge base*—to determine the *value* of candidate learning tasks, strategies, and tools as well as their possible combinations.

Kim et al. (2013) believe that intelligent learning that combines social learning with ubiquitous learning is a learner-centric and service-oriented education paradigm rather than focusing solely on the equipment use. At the 2010 *Korean Intelligent Learning Forum* (Noh 2011), the concept of intelligent learning was proposed as follows: First, it focused on humans and content, not devices. Secondly, it is an effective and intelligent tailor-made learning based on advanced IT infrastructure.

MEST (The Korean Ministry of Education, Science and Technology) defined *smart education* as *Self-directed, Motivated, Adaptive, Resource-enriched, and*

Technology-embedded education and learning systems. Their smart learning and education model so-called *S.M.A.R.T.* which consists of the following components:

S: Self-Directed means that the education system is progressing toward a self-learning system. Students' roles transit from knowledge adopters to knowledge creators. Teachers become learning facilitators.

M: Motivated means education becomes experience centered and involves learning by undertaking creative problem-solving and individualized assessment.

A: Adaptive means strengthening education system's flexibility and tailored learning for individuals' preferences and career prospects.

R: Resource-enriched means that Smart Learning adopts ample content based on open market, cloud education services from both public and private sectors. In other words, it expands the scope of learning resources to include collective intelligence—Social Learning.

T: Technology-embedded means that in a Smart Learning education environment, students can learn anywhere, any time through advance technologies.

Figure 11.2 shows the holistic concept of *SMART* education in Korea (Kim et al. 2013; Middleton 2015). It also stipulates learner-centric aspects and how it benefits from using smart technologies. The personalized and smart

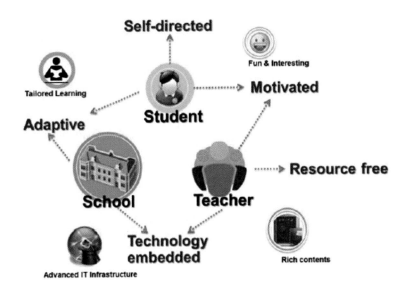

Fig. 11.2 Smart education proposed by Kim et al. (2013)

technologies make learners engage in learning and increase independence in a more open, connect, and augmented ways with distinctive ample contexts.

11.3 The Technology of Smart Education (TEL)

The basis of smart education is to use smart devices and smart technologies as a new educational paradigm (Kim et al. 2013). It is described as *technology enhanced learning* (TEL). TEL is used to provide flexibility in learning modes. They can be media or tools for content access (Daniel 2012), query, communication and collaboration, construction (Bruce and Levin 1997), expression (Goodman 2003), and evaluation (Meyer and Latham 2008).

Personalized technology has promoted mobile devices development. Mobile learning has become the main TEL mode. It focuses on learners' mobility compared with the static traditional education style (Hwang et al. 2008).

Many studies have begun recently to focus on the importance of real activities that enable students to solve real-world problems (Hwang et al. 2008). It is very important to design a combination of real and virtual learning environments to put students in a real learning environment. Seamless learning overlaps with some aspects of mobile learning and ubiquitous learning. It is described as a one-to-one TEL model for learners to learn across time and locations. And it is easy to convert learning scenarios from one to another through smart personal devices to cover formal and informal learning, personal, and social learning (Chan et al. 2006).

Cloud computing, learning analysis, big data, IoT, wearable technology, and other smart technologies have promoted the emergence of smart education. Cloud computing, learning analysis, and big data focus on how to capture, analyze, and guide learning data to improve learning and teaching, and support personalized and adaptive learning development (Mayer-Schönberger and Cukier 2013; Picciano 2012). Multifunctional learning technology, the learning platform can respond to individual learner data and adjust teaching resources. It can use aggregated data across large-scale learners to gain an in-depth understanding of course design (NMC 2015).

The Internet of Things and wearable technologies also support contextual learning and seamless learning development. IoT connects individuals, objects, and devices. The smart devices that learners carry with them can benefit from various relevant information that the surrounding environment advertises to them (NMC 2015). Wearable technology can integrate location

Fig. 11.3 IoT in a smart classroom

information, motion recording, social media interaction, and visual reality tools with learning. Figure 11.3 illustrates the most frequently used IoT in smart classrooms including: interactive whiteboards, e-books, tablets, 3D printers, student smart cards, sound sensors, security cameras, temperature sensors, light sensors, smart HVAC systems, attendance tracking systems, and wireless door locks, etc.

11.4 Framework of Smart Education

Reference to smart education projects from different countries and the meaning of smart, a Smart Education Framework with three major components is proposed by Zhu et al. (2016), illustrated in Fig. 11.4:

1. Smart learners.
2. Smart pedagogy.
3. Smart learning environments.

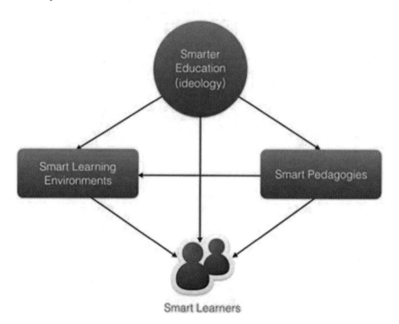

Fig. 11.4 Framework of Smart Education (Zhu et al. 2016) (CC BY 4.0)

The main theme of smart education is to create intelligent environments using smart technologies, so that smart pedagogies can facilitate personalized learning services and foster learners for better value orientation, stronger conduct ability, and talents of wisdom (Zhu et al. 2016).

These three components are interrelated. Smart education emphasizes the idea of pursuing better and/or *smarter* education to solve the needs of smart teaching methods as methodological issues and smart learning environments as technical issues to advance educational goals and train smart learners. Smart education may affect the smart environment seriously. Smart teaching methods and smart environments support smart learners' development.

11.4.1   Smart Learners

Learning is defined as the process of acquiring *competence* and *understanding* traditionally. It brings a new ability to do things and an understanding of unknown things. Ability refers to having specific skills and knowledge. The twenty-first century requires individuals to possess skills and abilities to operate and live effectively. The purpose of smart education is to train smart learners to meet these needs.

There are many organizations that develop twenty-first century skills independently. The *Organization for Economic Cooperation and Development (OECD)* has organized 10 skills in the twenty-first century into four categories, including ways of thinking, working tools, working methods, and world lifestyles (Ananiadou and Claro 2009). *Partnership for twenty-first century* (P21 2015) proposes a framework for learning in the twenty-first century, indicating that students should master the following knowledge and skills: key themes in twenty-first century; learning and innovation capabilities; information, media, and technical skills; life and career skill. *The North Central Regional Education Laboratory (NCREL)* proposed that literacy, creative thinking, effective communication, and high productivity in the digital age constitute the twenty-first century skills (Burkhardt et al. 2003).

There are four ability levels in smart education that students should master to meet the needs of modern society based on these studies. These abilities are basic knowledge and core skills, comprehensive abilities, personalized expertise, and collective intelligence:

1. Basic knowledge and core skills refer to knowledge and skills in core subjects such as STEM, reading, writing, art, etc. Mastery of these core subjects is essential to students' success (P21 2015). Jenkins (2009) also considered that the reading, writing, and mathematics are core capabilities of the twenty-first century.
2. Comprehensive abilities refer to abilities to think critically and solve real-world problem. Most of the twenty-first century skills frameworks raise the demands of thinking ways (Ananiadou and Claro 2009; Burkhardt et al. 2003; P21 2015). These abilities let the student use appropriate reasoning and comprehensive thinking in different complex situations. Students should solve different problems and produce better solutions based on analysis to make judgments and decisions.
3. Personalized expertise demands students to master information and technology literacy, creativity, and innovation skills. Information and technology literacy demand students to master ICT skills using different applications and combining cognitive abilities or higher-order thinking skills for learning (Ananiadou and Claro 2009). Creativity and innovation skills demand students to think and work creatively with others.
4. Collective intelligence refers to knowledge built up by a group of individuals via communication and collaboration. Students need to reflect about the ways to share and transmit results or outputs to others after the work with information and knowledge (Ananiadou and Claro 2009). They need to communicate in various ways clearly and effectively.

11.4.2 Smart Pedagogy

Learning methods become more and more flexible and efficient as technology advances. *Cognitive science* research shows that knowledge and skills are closely intertwined to enable learners to understand and put into practice. *Critical thinking* and learning skills are extremely important, but these skills cannot be taught independently, because some appropriate factual knowledge needs to be taught in a specific field and background (Ananiadou and Claro 2009). Using deliberate teaching or related learning strategies can develop knowledge and skills for learners.

Students acquire basic knowledge and core skills in the classroom. Every student's goals and processes are the same. But students with different backgrounds have different preparation levels, interests, and learning characteristics. Every student should receive an equal education, supplemented by content and performance standards that enhance understanding (Tomlinson and McTighe 2006). The process should be customized according to learning requirements, background, interests, preferences, etc. (Sampson et al. 2002). Interest-driven personalized learning can cultivate intrinsic motivation and promote students' personalized professional knowledge (Atkins et al. 2010).

In addition, whether it is in the classroom or online, usually need a group or team to achieve a common goal. The collaborative process enables students to develop comprehensive critical thinking and problem-solving skills (Stahl et al. 2006). Cooperative teams can retain knowledge by sharing information and discussions at a higher level of thinking to be responsible for their own learning (Totten et al. 1991).

Intelligence is the ability to get things done. Sternberg (1999) describes three basic aspects of successful intelligence including: analytical thinking, creative thinking, and practical applications. We provide students with the ability to solve problems, make decisions, creative thinking, and interest-driven learning so that they can generate intelligence. This is like transfer learning, or what we have learned to apply to other different related conditions in a specific situation (Barnett and Ceci 2002). Learning is a generating process. In this process, the learner is an active recipient of information, and he is committed to construct meaningful information found in the environment. Generative learning enables learners to apply what they have learned to a variety of relevant future situations (Fiorella and Mayer 2015).

The four instructional strategies proposed by Zhu et al. (2016), to foster learners' performance is shown in Fig. 11.5. These strategies include class-based differentiated instruction, group-based collaborative learning,

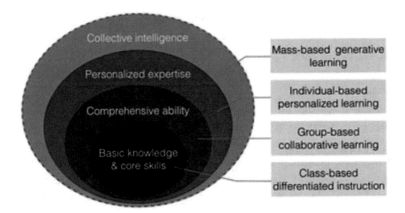

Fig. 11.5 Four-tier architecture of smart pedagogies (Zhu et al. 2016) (CC BY 4.0)

individual-based personalized learning (interest-driven predominantly), and mass-based generative learning (through online interactions predominantly).

11.4.3 Smart Learning Environments

Individuals criticize the traditional learning paradigm for being too artificial, rigid, and unresponsive to today's social needs (Kinshuk and Graf 2012). New technologies and new teaching methods developed in the digital age have attracted learners and promoted the learning of common phenomena. Piccoli et al. (2001) define and expands learning environment dimensions including space, place, time, technology, control mechanisms, and interactions. Therefore, it is possible to design a new learning environment technically and pedagogically.

From the technical point of view, *ambient intelligence (AmI)* is developing as a new research paradigm rapidly (Shadbolt 2003). In AmI environment, the device allows individuals to perform daily activities and tasks in a relaxed and natural manner from the network. It can interact and communicate independently without manual coordination, without making decisions based on a range of factors (including personal preferences and others in the surrounding area). Most students are digital natives who use smart mobile devices and digital resources for communication, learning, and entertainment every day (Bennett et al. 2008).

From the teaching perspective, the slanted analysis method enables institutions to support learners' progress and enrich personalized learning (Siemens and Long 2011). The goal is to monitor the learning process and use data analysis to predict students' future performance to discover potential problems (Zhu and Shen 2013). Teachers can provide information feedback through virtualized dashboards and learning analysis and obtain an overall view of learner activities related to peers or other participants in the learning experience.

The technically supported intelligent learning environment should not only enable learners to use digital resources and interact with ubiquitous learning systems, but also provide the necessary guidance, support tools or suggestions in the right place, time, and form actively (Hwang 2014). There are many hardware and software technologies that can support reinforcement learning. The hardware includes tangible objects, such as interactive whiteboards, smartwatches, e-bags, mobile phones, wearable devices, smart devices, sensors using pervasive computing, cloud computing, environmental intelligence, IoT technology, etc. The software includes various learning systems, learning tools, online resources, educational games using social networks, learning analysis, visualization, and virtual reality.

The goal of smart learning environments is to provide abundant, accurate, personalized, and seamless (formal and informal) learning experience using learning analytics. Ten key features of smart learning environments are as follows (Zhu et al. 2016):

1. Location-Aware: Sense learner's location in real time.
2. Context-Aware: Explore different scenarios and activities information.
3. Socially Aware: Sense social relationship.
4. Interoperability: Set standard between different resources, services, and platform.
5. Seamless Connection: Provide continuous service when any device connects.
6. Adaptability: Promote learning resource according to learning access, preference, and demand.
7. Ubiquitous: Predict learner's demand until express explicitly, provide a visual and transparent way to access learning resources and service to the learner.
8. Whole Record: Record learning path data to mine and analyze deeply, then give a reasonable assessment, suggestion, and promote on demand service.
9. Natural Interaction: Transfer multimodal interactions such as position and facial expression recognition.
10. High Engagement: Involve in multidirectional interaction learning experience.

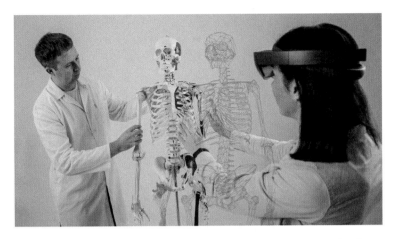

Fig. 11.6 AR-VR technology in smart education of medical training (Tuchong 2020)

Figure 11.6 shows VR-AR technology application in medical training.

11.5 3D Holographic AI Teacher

As we can see, the existing Smart Education comprises of state-of-the-art technology covering IoT, wireless communication, VR, and AR. With the advance AI technologies studied in this book, future Smart Education should focus on how different AI technologies such as computer vision, machine learning, data mining, NLP, and ontological search engine can be integrated together to provide a comprehensive solution for Smart Education.

In the following two sections, we will introduce two innovative AI-oriented Smart Education applications: 3D Holographic AI Teacher and Language Chatbot Tutor which are part of UIC iCampus research project applications.

Imagine one day we attend a lecture on AI and neural networks. It is possible, technology speaking, to implement AR-VR with the integration of AR technology; to demonstrate the biological structure of the brain wearing VR-AR glasses as shown in Fig. 11.7. How about we see 3D AR graphic without glasses? Or, even the lecturer is only a 3D software agent image?

The answer is absolutely yes. With the *Holographic Technology*, it is technically feasible to visualize 3D AR images with the naked eye, akin to what you see in many sci-fi movies. The challenges nowadays are: whether we can incorporate with various AI technologies such as machine learning, NLP, and ontological knowledge base to implement a 3D holographic AI teacher—a software robot with ontological knowledge-based on a particular knowledge domain (e.g. AI in our case) with the appearance of a 3D holographic image

Fig. 11.7 Holographic AR teacher (Tuchong 2020)

(of a real teacher) to deliver lectures and interact with students ubiquitously. Such technology becomes extremely useful at the time of writing this book when many of the author's colleagues are unable to conduct face-to-face teaching due to Covid-19.

11.6 Language Chatbot Tutor

Language learning is always a crucial task in education for foreign language(s) learning and oral practice. As revealed in Chap. 6 *Natural Language Processing (NLP)*, current NLP technology provides a feasible and practical solution to handle both Text-to-Voice and Voice-to-Text technology for frequently used languages with so-called *true human voice* synthesis. Today challenges are focused on how to integrate with other AI technology such as machine learning, data mining and ontological knowledge base with search engine to implement a comprehensive Smart Education solution for language learning and practice. *Language Chatbot Tutor* is one of the major AI research topics in UIC iCampus research project. AI technology integration such as NLP, machine learning (particularly in reinforcement learning technology), ontological knowledge base and search engine technology. Language Chatbot Tutor aims at the implementation of a comprehensive AI-based English language chatbot agent that can be installed in any mobile device for teaching and learning foreign language(s) such as English.

In terms of AI function, *English Language Chatbot* provides three levels of English learning: (1) Syntactic Level; (2) Semantic Level; and (3) Pragmatic Level which corresponds to the three knowledge levels in natural language. In Syntactic level, the user requires to read text passage (or dialog) shown by the chatbot. It will compare it with the sample voice soundtrack and advise student pronunciation as shown in Fig. 11.8. Technically, is alike to have an accompanied language tutor to correct pronunciation mistakes in daily dialog. In Semantic Level, the chatbot tutor will base on a specific scenario, e.g. inside a restaurant to teach students converse, pronounce, language styles, and meanings. In the Pragmatic Level, also the highest level, the language chatbot tutor will have the so-called *free conversation* with students, check and response language pronunciation and correct use of the language. In fact, it is also a major challenge for AI chatbot.

11.7 Case Study: A New Era of Smart Education

In this chapter, we studied various aspects of Smart Education ranging from its framework to different models implemented at countries worldwide.

The integration of different AI technologies such as machine learning, data mining, computer vision, NLP, and ontological knowledge base system and search engine with *smart education* is definitely a future trend. As a student, which aspect(s) of academic life and study will be affected and/or benefited by such technologies and changes?

Form a group of 4–5 students:

1. Explore and discuss which aspects of academic life may change in this new era of smart education, such as attendance, study, assignment, and examination, etc.
2. For each aspect of teaching and learning activity, discuss how it can be changed and what AI technologies are involved.

Fig. 11.8 Language chatbot tutor

3. For each aspect of teaching and leaning activity, discuss how we can get benefited from these smart education technologies.
4. What are the potential threats or problems that might arise from these smart education technologies or systems? (Fig. 11.9)

Fig. 11.9 Smart education with AI technology (Tuchong 2020)

11.8 Conclusion

In this chapter, we explore one of the foremost applications of AI in our daily activities—Smart Education. Education today is not only restricted to students and youngsters. Every one of us is in the course of continuous education in the sense that we all need to improve and upgrade our knowledge even if we have completed studies and started careers.

As some might worry: *Will AI and robotics replace our work, says as professional educators?* The similar kind of worry occurred when we started to use a computer back in 1970s, as many were worried about the work will be replaced by computers.

The truth is, there might be some changes in shifting our work nature alike with the use of computers, some tedious and so-called mechanical work will be replaced by computers allowing us to work on more creative and non-tedious work. The same logic applies to situations today. AI adoption in various industries will somehow replace some jobs and tasks that can be done by AI machines or systems more effectively. But again, we can improve ourselves and work on more creative and knowledge-oriented work. That is why Smart Education is significant. The importance of education is not only for self-benefit, but also to others and our next generations. Thus, the importance of Smart Education is to use the latest AI and related technology effectively so that it can benefit more people worldwide. This is the sole nature of Smart Education.

References

Ananiadou, K., & Claro, M. (2009). 21st century skills and competences for new millennium learners in OECD countries. *OECD Education Working Papers, 41.* https://doi.org/10.1787/19939019.

Atkins, D. E. (2010). Transforming American education: Learning powered by technology. *Learning, 114.*

Barnett, S. M., & Ceci, S. J. (2002). When and where do we apply what we learn?: A taxonomy for far transfer. *Psychological Bulletin, 128*(4), 612.

Bennett, S., et al. (2008). The 'digital natives' debate: A critical review of the evidence. *British Journal of Educational Technology, 39*(5), 775–786.

Bruce, B. C., & Levin, J. A. (1997). Educational technology: Media for inquiry, communication, construction, and expression. *Journal of Educational Computing Research, 17*(1), 79–102.

Burkhardt, G., et al. (2003). 21st century skills: Literacy in the digital age. *North Central Regional Educational Laboratory.*

Chan, F. M. (2002). ICT in Malaysian schools: Policy and strategies. In *ICT in Education* (pp. 15–22).

Chan, T. W., et al. (2006). One-to-one technology-enhanced learning: An opportunity for global research collaboration. *Research and Practice in Technology Enhanced Learning, 1*(01), 3–29.

Choi, J. W., & Lee, Y. J. (2012). The status of SMART education in Korea. *World Conference on Educational Multimedia, Hypermedia and Telecommunications, 1,* 175–178.

Daniel, J. (2012). Making sense of MOOCs: musings in a maze of myth, paradox and possibility. *Journal of Interactive Media in Education, 3,* Art-18.

Fiorella, L., & Mayer, R. E. (2015). Eight Ways to Promote Generative Learning. *Educational Psychology Review, 2015,* 1–25.

Goodman, S. (2003). *Teaching youth media: A critical guide to literacy, video production and social change.* Teachers College Press.

Hua, M. T. A. (2012). Promises and threats: IN2015 Masterplan to pervasive computing in Singapore. *Science, Technology and Society, 17*(1), 37–56.

Hwang, G. J. (2014). Definition, framework and research issues of smart learning environments-a context-aware ubiquitous learning perspective. *Smart Learning Environments, 1*(1), 1–14. https://link.springer.com/article/10.1186/s40561-014-0004-5.

Hwang, G. J., et al. (2008). Criteria, strategies and research issues of context-aware ubiquitous learning. *Educational Technology & Society, 11*(2), 81–91.

ICT Transforming Education Project. (2019). *ICT Transforming Education in Africa UNESCO-KFIT project update,* January–March 2019. https://unesdoc.unesco.org/ark:/48223/pf0000367858.

IBM, Smart Education. (2012). Retrieved May 17, 2020, from https://www.ibm.com/smarterplanet/global/files/au__en_uk__cities__ibm_smarter_education_now.pdf.

Jenkins, H. (2009). *Confronting the challenges of participatory culture: Media education for the 21st century.* MIT Press.

Kim, T., et al. (2013). (2013) Evolution to smart learning in public education: A case study of Korean public education. In L. Tobias, R. Mikko, L. Mart, & T. Arthur (Eds.), *Open and social technologies for networked learning* (pp. 170–178). Berlin, Heidelberg: Springer.

Kinshuk, S., & Graf. (2012). *Ubiquitous learning.* Berlin, Heidelberg, New York: Springer Press.

Kobayashi, T., et al. (2017). An application framework for smart education system based on mobile and cloud systems. *IEICE Transactions on Information and Systems, E100.D*(10), 2399–2410.

Mäkelä, T., et al. (2018). Student participation in learning environment improvement: Analysis of a co-design project in a finnish upper secondary school. *Learning Environments Research, 21*(1), 19–41.

Mayer-Schönberger, V., & Cukier, K. (2013). *Big Data: A revolution that will transform how we live, work, and think.* Houghton Mifflin Harcourt.

Meyer, B. B., & Latham, N. (2008). Implementing electronic portfolios: Benefits, challenges, and suggestions. *Educause Quarterly, 31*(1), 34–41.

Middleton, A. (2015). Smart learning: Teaching and learning with smartphones and tablets in post compulsory education (Media-Enhanced Learning Special Interest Group and Sheffield Hallam University, 2015).

New Media Consortium. (2015). The NMC Horizon report: 2015 higher education edition, 2015 (pp. 1–50). Retrieved May 17, 2020, from https://files.eric.ed.gov/fulltext/ED559357.pdf.

New York Smart Schools Commission Report. (2014). Retrieved May 17, 2020, from https://www.governor.ny.gov/sites/governor.ny.gov/files/archive/governor_files/SmartSchoolsReport.pdf.

Noh, K. S. (2011). An exploratory study on concept and realization conditions of Smart Learning. *The Journal of Digital Policy & Management, 9*(2), 79–88.

Partnership for 21st Century Learning, P21 Framework Definitions. (2015). Retrieved May 17, 2020, from https://www.p21.org/storage/documents/docs/P21_Framework_Definitions_New_Logo_2015.pdf.

Picciano, A. G. (2012). The evolution of big data and learning analytics in American Higher Education. *Journal of Asynchronous Learning Networks, 16*(3), 9–20.

Piccoli, G., et al. (2001). Web-based virtual learning environments: A research framework and a preliminary assessment of effectiveness in basic IT skills training. *MIS Quarterly, 2001,* 401–426.

Sampson, D., et al. (2002). Personalised learning: Educational, technological and standardisation perspective. *Interactive Educational Multimedia, 4,* 24–39.

Shadbolt, N. (2003). From the editor in chief: Ambient intelligence. *IEEE Intelligent Systems, 18*(4), 2–3.

Siemens, G., & Long, P. (2011). Penetrating the fog: Analytics in learning and education. *EDUCAUSE Review, 46*(5), 30.

Singapore Advances Intelligent Nation Masterplan. (2010). *Journal of E-Governance, 33*(4), 177–177.

Stahl, G., et al. (2006). Computer-supported collaborative learning: An historical perspective. In *Cambridge handbook of the learning sciences* (pp. 409–426).

Sternberg, R. J. (1999). Successful intelligence: Finding a balance. *Trends Cognitive Science,* 3(11), 436–442.

Tomlinson, C. A., & McTighe, J. (2006). *Integrating differentiated instruction & understanding by design: Connecting content and kids.* Alexandria: ASCD.

Totten, S., et al. (1991). *Cooperative learning: A guide to research.* Garland.

Tuchong. (2020). *AR-VR technology in smart education of medical training.* Retrieved May 17, 2020, from https://stock.tuchong.com/image?imageId=918804924712 484974.

Zhu, Z. T., & Shen, D. M. (2013). Learning analytics: The science power of smart education. *E-education Research, 5,* 5–12.

Zhu, Z., et al. (2016). A research framework of smart education. *Smart Learning Environments, 3*(4), 1–17. https://slejournal.springeropen.com/articles/10.1186/s40561-016-0026-2.

12

Smart City

Smart cities are those who manage their resources efficiently. Traffic, public services and disaster response should be operated intelligently in order to minimize costs, reduce carbon emissions and increase performance.

Mr. Eduardo Paes (Former mayor of Rio de Janerio, born 1969)

Abstract Smart city becomes a popular topic in information and communication technology (ICT) era worldwide. In a simple explanation, a smart city is a place with the integration of latest 5G, Internet-of-Things (IoT), and AI technology, where networks and services are more flexible, efficient, and sustainable with the use of information, digital, and telecommunication technologies to improve operations and provide an entirely new horizon of living and working environment for its inhabitants. This chapter first examines its major components and infrastructure followed by a progress review on different countries in the past decades. Then, we study its four critical masses: smart transportation, smart energy, smart healthcare, and smart technology. After that, we examine three major supporting technologies: IoT, Big Data, and AI and several applications/systems: smart pole, smart house, and smart campus.

Smart city becomes a popular topic in information and communication technology (ICT) era worldwide. In a simple explanation, a smart city is a place with the integration of latest 5G, Internet of Things (IoT), and

R. S. T. Lee, *Artificial Intelligence in Daily Life*, https://doi.org/10.1007/978-981-15-7695-9_12

AI technology, where networks and services are more flexible, efficient, and sustainable with the use of information, digital, and telecommunication technologies; to improve operations and provide an entirely new horizon of living and working environment for its inhabitants.

In this chapter, we will first examine its major components and infrastructure followed by a progress review on different countries in the past decades. Then, we will study its four critical masses: smart transportation, smart energy, smart health care, and smart technology. After that, we will examine three major supporting technologies: IoT, Big Data, and AI and several applications/systems: smart pole, smart house, and smart campus.

12.1   What is a Smart City?

What is a *smart city* and why are there many discussions? ICTs has experienced explosive growth due to hardware and software designs advancement at different activities in last several years that led to use many terms such as cyberville, digital city, electronic city, flexicity, information city, telicity, wired city, and smart city. Smart city is the largest abstraction among labels used as it encompasses other labels used for cities. In a smart city, digital technologies translate into better public services and resources for inhabitants while reducing environmental impact (Celino and Kotoulas 2013).

One of the formal definitions of smart city: *A city connecting the physical infrastructure, the information technology infrastructure, the social infrastructure, and the business infrastructure to leverage the collective intelligence of the city* (Harrison et al. 2010).

Another formal and comprehensive definition for smart city is: *An innovative city that uses information and communication technologies (ICTs) and other means to improve quality of life, efficiency of urban operations and services, and competitiveness, while ensuring that it meets the needs of present and future generations with respect to economic, social, and environmental aspects* (Ahvenniemi et al. 2017).

A broad overview of components for a smart city is shown in Fig. 12.1. There are different combinations to make cities smart but need not be labeled as smart. The number of components depends on cost and technology availability.

The world population has increased significantly in recent decades and so has living standards expectation. It is predicted that around 70% of the world population will live in urban areas by 2050. Cities currently consume 75% of the world's resources and energy to generate 80% of greenhouse gasses

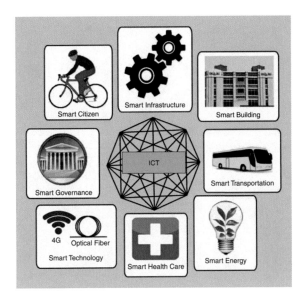

Fig. 12.1 Overview of smart cities (Mohanty et al. 2016)

approximately. There would be an enormous negative environmental impact in the next decades to compel the concept of smart cities a necessity. The creation of smart cities is a natural strategy to mitigate problems emerged from rapid urbanization and population growth. Despite the associative costs, smart cities once implemented, can reduce energy, water consumption, transport requirements, carbon emissions, and other waste.

Smart cities worldwide are diverse in terms of their characteristics, requirements, and components. In general, standards established by organizations such as the *International Organization for Standardization (ISO)* provide understood specifications globally to drive growth while ensuring quality, efficiency, and safety. IEEE has been developing standards for smart cities with different components, including smart grids, IoT, eHealth, and intelligent transportation systems (ITS). A specific example of the standard is ISO 37120 defines 100 city performance indicators, 46 core and 54 supporting that include economy, education, energy, and environment. These indicators can be used by city civic bodies to benchmark services performance and study best practices from other cities for comparison.

12.2 Major Components of Smart Cities

The characteristics of a smart city are shown in Fig. 12.2. There are nine

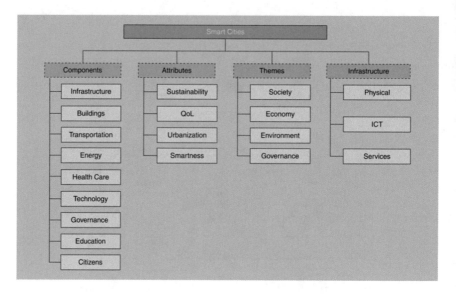

Fig. 12.2 Smart city framework (Mohanty et al. 2016)

components include smart infrastructure, smart buildings, smart transportation, smart energy, smart health care, smart technology, smart governance, smart education, and smart citizens. Some of these will be studied briefly in subsequent sections. Different smart cities have different smart components levels depending on their focuses (Mardacany 2014).

Attributes include sustainability, quality of life (QoL), urbanization, and smartness. Sustainability is related to city infrastructure and governance, energy and climate change, pollution and waste, social issues, economics, and health. QoL can be measured in terms of emotional and financial well-being of the citizens. Urbanization includes multiple aspects and indicators, such as technology, infrastructure, governance, and economics. Smartness is conceptualized as the ambition to improve economic, social, and environmental standards of the city and its inhabitants. Various commonly quoted aspects of city smartness include smart economy, smart people, smart governance, smart mobility, and smart living.

The core themes include society, economy, environment, and governance. Society theme signifies the city is for its inhabitants or the citizens. Economy theme signifies that the city is to thrive with continuous job growth and economic growth. Environment theme indicates that the city is to sustain its function and operation for current and future generations. Governance theme suggests that the city is robust in its ability to administer policies and combine other elements.

The infrastructure includes physical aspects, ICT, and services. Physical infrastructure is typically the non-smart component of smart cities but is the physical or structural entity comprises of buildings, roads, railway tracks, power supply lines, and water supply system. ICT infrastructure is the core smart component of the smart city that glues all other components, acting as the nerve center of smart city essentially. Service infrastructure is based on physical infrastructure that may have some ICT components. Examples include mass rapid transit system and smart grids.

12.3 Smart City Infrastructure

The smart infrastructure concept of a city is shown in Fig. 12.3 (Chourabi et al. 2012). It consists of physical components of the city, such as roads, buildings, bridges, that make the city and its inhabitants operate in a classic sense. However, in the context of smart cities, electrical, and digital systems are the backbone considered as its infrastructure. Other examples include rapid transit system, road network, railway network, communication system, traffic light system, street light system, water, gas and power

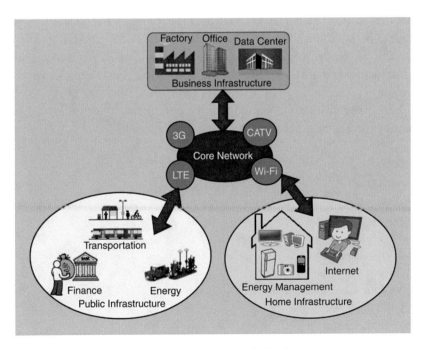

Fig. 12.3 Smart city infrastructure (Mohanty et al. 2016)

supply system, firefighting system, hospitals system, bridges, accommodation, hotels, office spaces, digital library, law enforcement, economy system, and waste management system. ICT infrastructure is the back end of smart infrastructure which makes the physical infrastructure *smart*. It is fundamental to smart cities construction that depends on factors related to its availability and performance. ICT infrastructure includes communication infrastructure such as fiber optics, Wi-Fi networks, wireless hotspots, as well as service-oriented information systems. It may have physical infrastructure, sensors, firmware, software, and middleware as its overall components. Smart infrastructure is more efficient, safe, secure, and fault-tolerant compared with classic infrastructure.

12.4   Progress of Smart Cities in Different Countries

Smart cities' major strategies and progresses worldwide are listed in the *Intelligent Community Forum* awards from 1999 to 2010 (IntelligentCommunity 2020). Songdo and Suwon (South Korea), Stockholm (Sweden), Gangnam District of Seoul (South Korea), Waterloo, Ontario (Canada), Taipei (Taiwan), Mitaka (Japan), Glasgow (Scotland, UK), Calgary (Alberta, Canada), Seoul (South Korea), New York City (US), LaGrange, Georgia (US), and Singapore were recognized for their efforts in developing broadband networks and e-services sustaining innovation ecosystems, growth, and inclusion.

While many cities have started implementing smart technologies, a few stood out further in development. These cities include:

- Barcelona, Spain
- Columbus, Ohio
- Dubai, United Arab Emirates
- Hong Kong, China
- Kansas City, Missouri
- London, England
- Melbourne, Australia
- New York City, New York
- Reykjavik, Iceland
- Shanghai, China
- Singapore
- Tokyo, Japan

- Toronto, Canada
- Vienna, Austria.

Most of the new smart city projects were concentrated in the Middle East and China but in 2018, Reykjavik and Toronto were listed alongside Tokyo and Singapore as some of the world's smartest cities.

Singapore uses sensors and IoT-enabled cameras to monitor public spaces cleanliness, crowd density and registered vehicles movement are often considered the of smart cities gold standard. These technologies help companies and residents monitor energy, water consumption, and waste production in real time. Singapore is also performing tests on autonomous vehicles including full-size robotic busses, as well as an elderly monitoring system, to ensure the health and well-being of senior citizens.

Kansas City, Mo., involves smart streetlights, interactive kiosks, free public Wi-Fi at 50 blocks along the city's two-mile streetcar route. Pedestrian hotspots, traffic flow, and vacant parking spaces information are available for the public through the city's data visualization app.

San Diego had installed 3,200 smart sensors to optimize traffic and parking, enhance public safety, and environmental awareness for residents in early 2017. Solar-to-electric charging stations are available to empower electric vehicle use, and connected cameras to help monitor traffic and pinpoint criminal activities.

Dubai, United Arab Emirates had used smart city technology for traffic routing, parking, infrastructure planning, and transportation. It also uses telemedicine and smart health care, as well as smart buildings, education, and tourism and utility services.

Barcelona, Spain had used smart transportation system and smart bus systems complemented by smart bus stops to provide free Wi-Fi, USB charging stations, and bus schedule updates for commuters. A bike-sharing program and smart parking app includes online payment is also available. The city also uses different kind of sensors to monitor temperature, humidity, rainfall, noise, and pollution levels.

The commitment to smart cities was firmly established by the Chinese government in its 12th Five-Year Plan issued in 2011. China started piloting national smart city development to encourage the latest technology use such as AI and IoTs to help traffic flow, improve law enforcement, and make public buildings more energy efficient in 2012. Three groups of cities were listed as national pilot projects and aim to nurture 100 new smart cities from 2016 to 2020, for urban planning and development. Shanghai, Beijing, Guangzhou, Xi'an, Yinchuan, and Hangzhou become notable examples of

Fig. 12.4 Hangzhou future smart city—*City Brain* (Tuchong 2020a)

older urban areas that have received smart city makeovers. Hangzhou, a city in East China, with a population of 9.47 million had a smart city system designed by Chinese tech giant Alibaba called *City Brain* was in use since 2016. *City Brain* is an AI system that uses big data and super-computing power to improve and fix traffic problems. By monitoring every vehicle in the city, *City Brain* has reportedly already reduced traffic congestion by 15%. Its aims to optimize traffic for the entire city, predict traffic congestion locations and even prevent traffic accidents by instituting preemptive traffic control and policing (Fig. 12.4).

12.5 Smart Transportation

Traditional transportation systems or facilities such as railway network, road transport, air, water transport existed a long time ago. Each operates in a specific type of transport system independently causing complex usage. Smart transportation, also known as an ITS includes various types of communication and navigation systems in vehicles, between vehicles (e.g. car to car), and between vehicles with fixed locations (e.g. car to infrastructure). ITS also covers railway, road, air, water transport systems and their interactions. A broad smart transportation concept is shown in Fig. 12.5. ITS has made it possible to construct global airway hubs, intercity railway networks, intelligent road networks, protected cycle routes, pedestrian paths, and integrated public transport for safe, efficient, reliable, and cost-effective transportation. ICT and real-time data processing maximize vehicle utilization in the

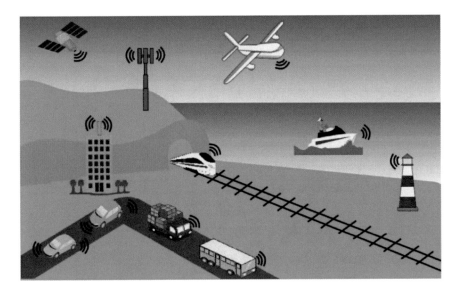

Fig. 12.5   Illustration of smart transportation (Mohanty et al. 2016)

system that allows passengers to select transportation options economically and effectively.

12.6   Smart Energy

What is smart energy? Smart energy is much broader than any of the previous concepts. It can be regarded as an *Internet of Energy* model. This model is based on one or more principles of smart power generation, power grids, storage, and consumption. Any traditional, clean, green, sustainable, and renewable energy along with ICT make smart energy in general. The major components and system framework are shown in Fig. 12.6.

Smart energy system consists of intelligent integration with decentralized sustainable energy sources, efficient distribution, and optimized power consumption. Smart energy includes three independent building blocks that must be adhered closely and communicated with each other to form a unified smart energy system effectively. Low-carbon generation, also known as green energy, photovoltaic, solar thermal, biogas, and wind energy are significant sustainable energy sources of the system. The use of smart infrastructure, grids, meters, pertinent ICT level utilization to collect energy consumption and share provider rates information is the second component of the system warrants efficient distribution. ICT can be used to control energy consumption levels for plug-in electric vehicles (PEVs), heating, ventilation,

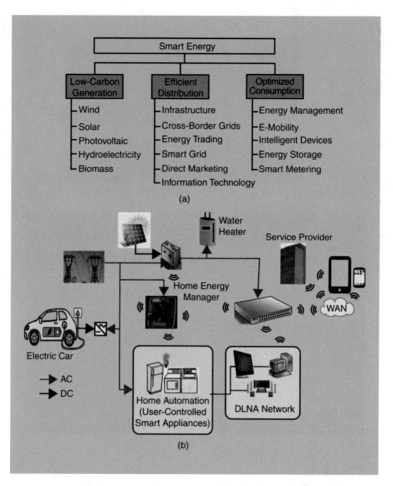

Fig. 12.6 Smart energy **a** components **b** system framework (Mohanty et al. 2016)

air conditioner, and household appliances. It can also be used to purchase energy from diverse sources including solar panels systems and wind turbine systems effectively. Optimized consumption of the system is the third key component of the system. Smart metering, energy storage, and management are keys to optimize energy consumption in a smart energy system.

12.7 Smart Health Care

Traditional health care is overwhelmed by the rapid growth of the population. Medical professional and equipment shortages cause huge difficulties to cope with pandemic disease outbreaks, and on some occasions, with inaccurate diagnosis. There are remote places where receiving adequate health care remains a distant dream. With limited resources and an ever-increasing demand, traditional health care needs to be intelligent, efficient, and sustainable. Smart health care can be conceptualized as several entities combination which includes traditional health care, smart biosensors, wearable devices, ICTs, and smart ambulance systems. The concept of smart health care is shown in Fig. 12.7.

The various components of smart health care include emerging on-body sensors, smart hospitals, and smart emergency response (Demirkan 2013). In smart hospitals, different operations mechanism such as ICTs, cloud computing, smartphone apps, and advanced data analysis techniques are used. Patient data can be available in real time and accessed by medical professionals at various offices in a smart hospital or even various hospitals in the same city or in different cities without time lost. Similarly, different specialists can study a patient's condition to make judgments with corresponding medication.

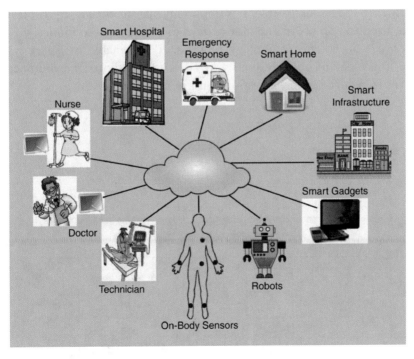

Fig. 12.7 Major components of smart health care (Mohanty et al. 2016)

Telemedicine is a specific example and a subset of smart health care. ICTs are used to provide clinical health care from long distance or remote locations. This approach is particularly useful where health-care services are uneasily accessible. It eliminates distance barriers and improves medical care access. Telemedicine is envisioned to provide critical care in emergency situations to save lives. Another example applies in assisted living for the elderly to enjoy independence and quality of life with minimal skilled nursing care but are accessible when needed.

12.8   Smart Technology

Smart technology is key for the design, implementation, and operation of smart cities (Maeda 2012) shown in Fig. 12.8. A diversity of components include infrastructure, physical structure, buildings, electrical infrastructure, electronics, communication infrastructure, information technology infrastructure, and software. The design and operation challenges are a good mix of smart technologies so that the smart cities are not over smart yet sufficiently smart to be sustainable. It is important that the deploying cost is not a major overhead on tax revenue for citizens to become an economically viable option as science and technology progress. A sustainable transport system is an important technology for smart cities to transport large numbers of citizens to reduce traffic congestion and greenhouse emissions.

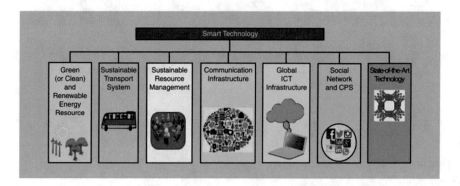

Fig. 12.8   Major components of smart technology (Mohanty et al. 2016)

Smart communication technology and ICT are important infrastructure to include fiber optics, citywide Wi-Fi, near-field communication (NFC), and Bluetooth. Citywide Wi-Fi can offer basic services such as ride-hailing easier. NFC can revolutionize the way debit cards are used when it becomes a cashless society. Cyber-physical systems (CPS) are computation, networking, and physical entities integrations like IoT are essential in making smart physical entities. Social networks and short-message services have already created communications mechanisms to avail utilities in smart cities efficiently.

There are a variety of state-of-the-art technologies used to make cities smart. Wi-Fi and NFC are considered part of this trend but there are many other forms. For instance, a smart meter can be used to measure and record utilities consumption and communicate the information for monitoring and billing to central facilities. Another state-of-the-art technology is electronic cards or smart cards that contain a unique encrypted identifier to allow the owner access to a range of services without setting up multiple accounts. In addition, a secure digital camera's network can be an effective solution to secure and copyright images or video communication in IoT use in smart health care and smart transport.

12.9 IoT and Smart City

IoT is the core of smart city implementation (Zanella et al. 2014; Rossetti 2015). It is the smart cities technical backbone as shown in Fig. 12.9. Smart cities need to have three key features IoT can provide: intelligence, interconnection, and instrumentation.

Smartphones, meters, sensors, and RFID form an IoT framework in smart cities. IoT framework consists of various components including electronics, sensors, networks, firmware, and software. It is the physical objects (called Things) interconnected network that include computers, smartphones, sensors, actuators, wearable devices, homes, buildings, structures, vehicles, and energy systems. IoT ensures different communication system types and applications to provide smart, reliable, and secure services increasingly. A large variety of sensors include RFID, IR, and GPS connect buildings, infrastructure, transport, networks, and utilities through ICT to process information exchange, communication, intelligent recognition, location determination, tracking, monitoring, pollution control, and identity management performed by IoT framework. Thus, CPS can be brought to the discussion in relation to IoT. It is difficult to distinguish between CPS and IoT based on the available literature since CPS is a much larger entity

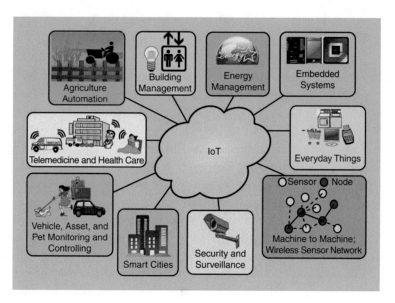

Fig. 12.9 IoT in smart cities (Mohanty et al. 2016)

than IoT; in other words, IoT is a network/communication CPS subset. It is the IoT implementation in a physical system that leads to CPS.

12.10 Big Data and Smart City

Big Data (BD) refers to a collection of large and complex data sets that would be difficult to process using regular database management tools or traditional data processing applications in general. IoT, BD, and smart cities are strongly interrelated, as one needs the other two. The urban data that are tagged in space and time generated in smart cities can be BD. BD may be generated from large sensors, databases, e-mails, websites, and social media collection shown in Fig. 12.10 (Jara et al. 2014; Harris 2014).

It is estimated that the proliferation of sensors, social networks, web pages, image and video applications, and mobile devices generate more than 2.5 quintillion bytes per day. BD has multiple challenges that include visualization, mining, analysis, capture, storage, search, and sharing. BD requires new processing approaches to enhance decision making, insight discovery for optimization. Sophisticated data analysis mechanisms are necessary to search and extract valuable patterns and knowledge from IoT BD. BD has several important characteristics that include complexity, volume, variety, variability, and veracity. It has three abstraction level types: enterprise data, public data,

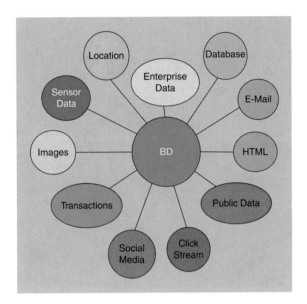

Fig. 12.10 Big Data and data mining in smart cities (Mohanty et al. 2016)

and transactions shown in Fig. 12.10. BD examples include atmospheric data, call detail records, genomic data, e-commerce data, Internet search indexing, medical records, military surveillance, photography archives, RFID data, sensor network data, social network data, video archives, and weblogs. BD must be processed with advanced analytics, algorithmic methods, and tools to store and retrieve meaningful information when needed. The storage cost is inexpensive due to affordable storage facilities.

12.11 AI and Smart City

Smart cities are already using AI at a greater pace in the new century. It is important for cities to identify using AI for a specific goal in its implementation process and not because is the newest trend in the technology market available (Fig. 12.11).

Building a smart city is not a one-day business, neither is an individual's work nor organization. It requires the collaboration of many strategic partners, leaders, and even citizens. Any AI-based smart city with AI, IoT, and ICT integration requires the following:

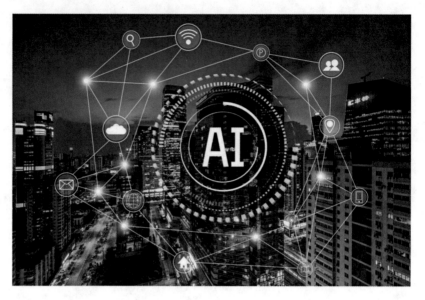

Fig. 12.11 AI in smart cities (Tuchong 2020b)

- A network of *smart IoT* (sensors, cameras, actuators, and so on) to gather data
- Field (cloud) gateways to gather data from low power IoT devices, store, and forward it securely to the cloud
- Stream data processor to aggregate numerous data streams and distribute them to a data lake and control applications
- A data lake to store all raw data, even the ones that seem to be no value
- A data warehouse to clean and structure the collected data
- Tools to analyze and visualize data collected by sensors
- AI algorithms and techniques for city automation services based on long-term data analysis and find ways to improve control applications performance
- Control applications to dispatch commands to IoT actuators
- User applications to connect *smart IoT* and citizens.

Figure 12.11 shows an AI, IoT and ICT concept integration in future smart cities. In the next sections, we will explore several remarkable applications for smart cities: smart pole, smart house, and smart campus.

12.12   Smart Pole

The humble lamppost is receiving a lot of attention these days from cities worldwide. It is recognized that the poles can do a lot more apart from streetlights uphold aloft. They can be *Smart Poles*.

A smart pole is a multifunctional pole designed host to technologies and utilities such as *access network* (small cell, Wi-Fi), *IoT* (detectors, sensors, and communication), *green energy* (solar power, efficient LED lights), *sensors, advertisement, citizen interactions,* etc., in general, as shown in Fig. 12.12.

Intelligent, multifunctional, or smart light poles help to solve many urban problems to incorporate software controls, electronics, and sensors to receive and transmit data. They can improve traffic management and parking through real-time data, reducing congestion and emissions. Intelligent poles can also monitor air quality, detect, and notify officials about street flooding or turned into charging stations for electric vehicles and support other smart city applications, such as the surge of driving navigation systems as they are equipped with internet services.

The smart pole market is expected to witness significant growth by these factors. Further, government worldwide are increasingly taking initiatives to implement smart city plan at disconnected regions. These enable end users to connect with developed regions of the country and stay updated. The installation is also expected to extend growth opportunities for industries in backward regions of various countries.

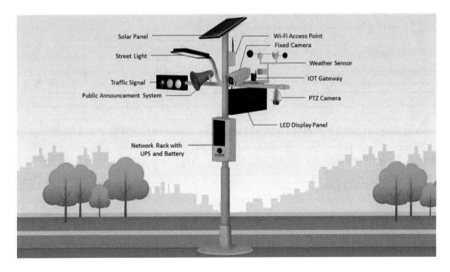

Fig. 12.12 Smart pole

12.13   Smart House

What is Smart House and How It Works?

A *smart house* is an accommodation that has highly advanced, automated systems to control and monitor functions of a house such as lighting, temperature control, multi-media, security, window and door operations, air quality, or any other necessity task performed by a resident.

A smart house appears *intelligent* because its computer systems can monitor many aspects of daily activities. For example, the refrigerator may be able to take inventory of its contents, suggest menus and shopping lists, recommend healthy alternatives, and order groceries. The smart home systems may even ensure a cat litter box or house plans are cleaned and watered regularly.

The smart house idea may sound like something of a Hollywood movie. In fact, a Disney movie titled *smart house* did present comical antics of an American family won a *house of the future* with an android maid who caused havoc in 1999. Other science fiction films show visions seem improbable.

However, smart home technology is real and is becoming increasingly sophisticated. Coded signals are sent through home's wiring (or sent wirelessly) to switches and outlets programmed to operate appliances and electronic devices in every part of the premises. Home automation is exceedingly useful for the elderly, individuals with physical or cognitive impairments and disabled who wish to live independently.

Pros Versus Cons of Smart House

Smart house technology system installation provides homeowners convenience. The devices are interconnected and accessed by one central point—a smartphone, tablet, laptop, game console, or another networked device. Door locks, televisions, thermostats, home monitors, cameras, lights, and even appliances such as the refrigerator can be controlled by one home automation system. User can create certain changes at times scheduled (Fig. 12.13).

Smart home appliances come with self-learning skills so they can learn the homeowner's schedules and adjustments as needed. Smart homes enabled energy control for homeowners to reduce use and benefit from related cost savings. Some home automation systems are equipped with motion detection to alert the homeowner when away, and contact the police or fire department in case of emergency situations. Users can receive notifications and updates about their accommodation. For instance, smart doorbells allow homeowners

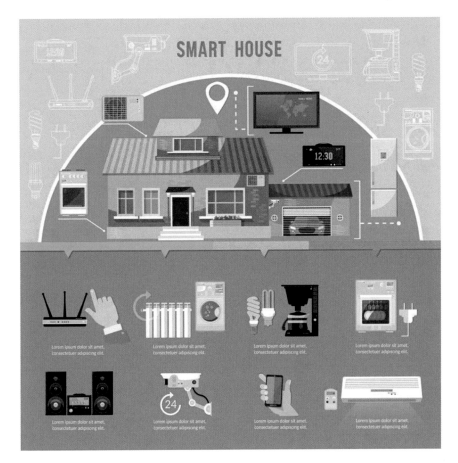

Fig. 12.13 Smart house (Tuchong 2020c)

to see and communicate with individuals who come to their doors when not at home. Once connected, services such as smart doorbell, security system, and appliances are all part of IoT technology, a physical objects network to gather and share electronic information.

While the smart house technology offer cost savings and convenience but they also post challenges. Security risks and bugs continue to plague makers and users. Adept hackers, for example, accessed to a smart home's internet-enabled appliances. There was a botnet called Mirai infiltrated interconnected devices of DVRs, cameras, and routers to bring down a host of major websites through a denial of service attack, also known as a DDoS attack in October 2016. Measures to mitigate the risks of such attacks include protecting smart appliances and devices with strong passwords, using encryption when available and only connecting trusted devices to one's network. As noted above, the costs of installing smart technology can cost around a few thousand for a wireless system to tens of thousands for a hardwired system. It's an expensive price, especially when there may be a steep learning curve for everyone in the household to get used to the system.

12.14   Smart Campus

What is Smart Campus?

Smart Campus is a term used to describe educational institutions that use next-generation technologies woven seamlessly within a well-architected infrastructure. This enables a *digitally connected* institution to enhance campus experience, drives operational efficiency, and provides education in a manner that all can access. It creates proactive interactions to foster positive outcomes.

A smart campus is more than a system or collection of applications, platforms, or infrastructures in silos around campuses. It requires a strategic framework that strengthen a connected ecosystem and creates a new, exciting experience for all. Below are the capabilities within the framework to promote digital experience.

Smart Campus Framework

It uses next-generation technologies such as AI, machine learning, blockchain, facial recognition, smart sensors, and beacons—technologies that are strategically placed to trigger communications, take attendance automatically, derive venue analytics, automate processes, monitor and initiate workflows, and offer many more innovative practices that digital natives are accustomed to.

Deloitte Smart Campus framework (Deloitte 2019) consist of the following core modules:

- Smart students
- Smart classrooms and labs
- Smart teaching and research
- Smart housing and dining
- Smart mobility—campus mobility and safety
- Smart events—campus event logistics and services
- Smart campus operations.

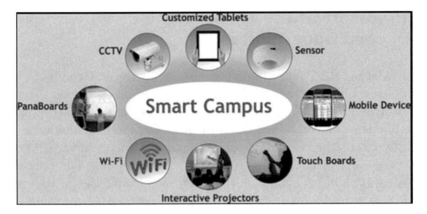

Fig. 12.14 Smart campus

Figure 12.14 shows major components and facilities in smart campus.

Benefits of a Smart Campus

Smart campus implementation enhances paradigm shifts to transform institutions to serve the needs for present and future campuses. It can improve three important factors: experience, efficiency, and education. It can reshape how students learn, what they learn, and how they interact with an institution. It used as a catalyst for the campuses transformation to address the future of learning and work while reshaping the experience. It can continue to serve the traditional campus as needed and adopt new approaches to serve digital native stakeholders expect with below benefits (Deloitte 2019):

- Cultivate a digitally connected experience
- Address cost pressures
- Elevate communication and awareness
- Provide sense of safety
- Impart a quality of life
- Foster inclusion and equity.

12.15   Smart Cities—Challenges and Opportunities

The challenges for building smart cities are diverse and complex. They include cost, efficiency, sustainability, communication, safety, and security as shown in Fig. 12.15. These design challenges are governed by factors including natural environment, government policy, social communities, and economy. Cost is the foremost factor for the smart campus. It includes design and operations costs. The design cost is a onetime cost. The operations cost is required to maintain a smart campus operation. Design cost needs to be small to make smart campus realization possible. At the same time, the same for operations cost will make it easier to operate in the long run. Cost optimization over the complete system life cycle can be a challenging problem. Operations efficiency is an important challenge as higher efficiency can reduce operational cost and improve sustainability. Cutting down carbon emissions and campus waste also enhance sustainability, efficiency, and reduce operations cost. They also need to be resilient to disasters and failures.

Disasters come from Mother Nature. Failures originate from a system for many reasons such as an ICT failure or a power failure. Natural disasters can also lead to different components of failures. Any design needs to take these potentials into consideration so the campus can recover quickly when they occur. The design and operations cost will be affected by these challenges. Smart campus uses many smart components including ICT, sensors, and IoTs to process and store large data volumes to operate, so information and infrastructure security is also important to design challenge. Above all, inhabitant safety is paramount. These will all drive to increase the design and operations budgets.

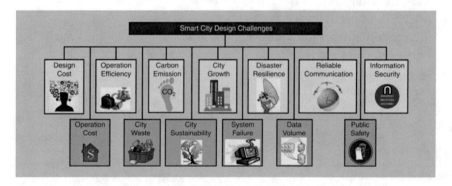

Fig. 12.15   New challenges and opportunities in smart cities (Mohanty et al. 2016)

12.16 Case Study: How Smart Cities Reshape Our Daily Activities?

In this chapter, we have discussed different aspects of Smart City and its supporting technology: IoT, big data, and cloud technology, AI and technology.

As mentioned, smart city projects implemented in the past decades were mainly focused on ICT with internet integration (especially IoT), computing and communication technology. In the new AI era, big data and 5G technologies, the integration of different AI technologies such as machine learning, data mining, computer vision, NLP, and ontological knowledge base system and search engine with Smart City is the future trend. As a smart city citizen, what aspect(s) of your daily activities will be affected and/or benefited by such technology?

Form a group of 4–5 students:

1. Explore and discuss daily activities aspects might be changed in this new era.
2. For each aspect of daily activities, discuss how they can be changed and what AI or related technologies are involved.
3. For each aspect of daily activities, discuss how you can benefit from these smart city technologies.
4. What are the potential threats or problems that may arise from these smart city technologies or systems? (Fig. 12.16)

12.17 Conclusion

In this chapter, we study one of the foremost smart technology that might affect our daily activities—*smart city*. The definition of a smart city and supporting technology being applied are mostly changed owing to the advancement of technology. In the past decades, smart city projects are mainly focused on ICT—*Information and Communication Technology* and integrated the latest computing and networking technology into IoT (Internet of Things) to enhance communication and various control applications, devices, and systems in the smart city. With the new AI era, big data and various AI related technology, together with the advancement of the latest 5G technology, the scope and capability of various components of smart city have completely changed from smart *connection and control* to *truly intelligence and smart*, ranging from smart transportation with drone technology

Fig. 12.16 Smart city and us (Tuchong 2020d)

to smart robot in our household. AI and smart technology will surely be a part of our daily activities in the near future, or already materialized at some intelligent cities.

Smart City and related smart technology can implement almost anything that appeared only in science fiction in the past 50 years. We need to consider: How can we position ourselves with such vital changes? How can we use these technologies economically not only for self-benefits but also for others with hardship, livelihood, and tuition difficulties? How can we use these technologies economically to better preserve and protect our environment? All require thoughtful consideration.

We will explore three more important AI topics in our daily activities in the final part of this book: AI and self-consciousness, AI ethics, and future AI.

References

Ahvenniemi, H., et al. (2017). What are the differences between sustainable and smart cities? *Cities, 60,* 234–245.

Celino & Kotoulas, S. (2013). Smart cities. *IEEE Internet Computing, 17*(6), 8–11.

Chourabi, H., et al. (2012) Understanding smart cities: An integrative framework. *The 45th Hawaii International Conference on System Sciences,* 2289–2297. https://doi.org/10.1109/HICSS.2012.615.

Demirkan, H. (2013). A smart healthcare systems framework. *IT Professional, 15*(5), 38–45.

Deloitte. (2019). *Smart campuses—The next-generation campus.* Retrieved May 19, 2020, from https://www2.deloitte.com/us/en/pages/consulting/solutions/next-generation-smart-campus.html.

Harrison, C., et al. (2010). Foundations for smarter cities. *IBM Journal of Research and Development, 54*(4), 1–16.

Harris, S. (2014). Securing big data in our future intelligent cities. *IET Conference on Future Intelligent Cities, 2014*(15564), 8. https://doi.org/10.1049/ic.2014.0049

Intelligentcommunity. (2020). Retrieved May 19, 2020, from https://www.intelligentcommunity.org.

Jara, A. J., et al. (2014). Big Data in smart cities: From Poisson to human dynamics. In *The 28th International Conference on Advanced Information Networking and Applications Workshops* (pp. 785–790). doi:https://doi.org/10.1109/WAINA.2014.165

Maeda, A. (2012). *Technology innovations for smart cities. Symposium on VLSI Circuits (VLSIC),* 6–9. https://doi.org/10.1109/VLSIC.2012.6243763

Mardacany, E. (2014). Smart cities characteristics: Importance of built environment components. *IET Conference on Future Intelligent Cities, 2014*(15564), 4. https://doi.org/10.1049/ic.2014.0045

Mohanty, S. P., et al. (2016). Everything you wanted to know about smart cities: The internet of things is the backbone. *IEEE Consumer Electronics Magazine, 5*(3), 60–70.

Rossetti, R. J. F. (2015). Internet of Things (IoT) and smart cities. *Readings on Smart Cities, 1*(7).

Tuchong. (2020a). *Hangzhou future smart city—city brain.* Retrieved May 19, 2020, from https://stock.tuchong.com/image?imageId=475520345119260745.

Tuchong. (2020b). *AI in smart cities.* Retrieved May 19, 2020, from https://stock.tuchong.com/image?imageId=803266095386460178.

Tuchong. (2020c). *Smart house.* Retrieved May 19, 2020, from https://stock.tuchong.com/image?imageId=262139231589367889.

Tuchong. (2020d). *Smart city and US.* Retrieved May 19, 2020, from https://stock.tuchong.com/image?imageId=258625261146472586.

Zanella, A., et al. (2014). Internet of Things for smart cities. *IEEE Internet of Things Journal, 1*(1), 22–32.

Part IV

Beyond AI

13

AI and Self-consciousness

A computer would deserve to be called intelligent if it could deceive a human into believing that it was human.
Sir Alan Turing (1912–1954)

Abstract This chapter studies one of AI's most challenging topics—the possibility of self-consciousness and self-aware robots (or AI systems). Since Turing Test to the latest AI development, we have never ceased to explore human intelligence, and the creation of intelligent systems and robotics. With the advancement of technology, almost all core AI functionalities including machine learning, computer vision, data mining, natural language processing, and agent ontology have been examined immensely with significant success and applied to our daily activities. Many AI scientists believe it is the time to explore this ultimate question—robot consciousness. This chapter begins with consciousness concepts and machine consciousness in neuroscience disciplines brief literature review to current AI and machine learning R&D. Next, we explore machine consciousness typical approach—the Good Old-Fashioned Artificial Consciousness (GOFAC) which consists of five major components: (1) Functionalism; (2) Information integration; (3) Embodiment; (4) Enaction, and (5) Cognitive mechanisms. Lastly, we conclude AI and machine consciousness study, outstanding issues; problems to approach in order to design and build a true self-consciousness and self-awareness robot.

R. S. T. Lee, *Artificial Intelligence in Daily Life*,
https://doi.org/10.1007/978-981-15-7695-9_13

This famous quotation from Sir Alan Turing forms the basis of *Turing Test* who also coined the start of AI research from 1950, to present. The questions are: *Does a machine which can deceive human to believe that it was human really intelligence? Does the machine really know what it is talking about? Does the machine really know it is a machine?*.

In this chapter, we will explore these ultimate questions and challenges—AI, *self-consciousness,* and *self-awareness.* These are in other words, to explore how *intelligent robots* (or systems) can have so-called *subjective experiences.* We will first introduce consciousness concepts and ideas with a brief literature review of machine consciousness in neuroscience disciplines to current AI R&D and machine learning. Then, we explore the typical approach on machine consciousness—the *Good Old-Fashioned Artificial Consciousness (GOFAC)* which consists of five major components: *(1) Functionalism; (2) Information integration; (3) Embodiment; (4) Enaction, and (5) Cognitive mechanisms.* Finally, we conclude the study of AI and machine consciousness, outstanding issues and problems to address in order to design and build a true self-consciousness and self-awareness robot.

13.1   Consciousness in Machine

We all believe that *consciousness* exists among ourselves per René Descartes' famous quote (Cottingham 2017): *Cogito Ergo Sum (I think; therefore, I am).* Although we cannot be certain about anything that happens among us. But at least, we know and aware of our own existence—*self-consciousness* and *self-awareness* (Cosmin 2017).

For many years, the definition of *consciousness* is still an open question. Therefore, discussing robot consciousness will be problematic: Raoult and Yampolskiy (2015) reviewed 21 literature to evaluate the consciousness of machines and robots. The same is true for other complex concepts such as *intelligence.* It is worth noting that Legg and Hutter (2007), reviewed more than 70 different definitions of *intelligence.* The fact is that there is no consensus on what *intelligence* is, which does not prevent researchers from talking about *artificial intelligence. Consciousness* is an important research topic in neuroscience: Dehaene (2014), summarizes the years of research on human consciousness; see also Tononi (2012) and Damasio (2010). It is worth noting that neuroscientists dedicated to consciousness have seriously considered the possibility that robots may have consciousness in the near future (Pandey 2018; Hildt 2019; Chella et al. 2020; Chatila et al. 2018).

From neuroscience perspective, Grossberg (2017), summarizes the results of years of research on brain resonance and proposes a set of models described by differential equations, which capture some of the main aspects of *consciousness*. Perlovsky (2016) pointed out that a new *psychophysics* is needed to find the basic laws of the physical world, including *emotions*. Some psychophysics start to develop mathematical theories to explain empirical evidence about emotions and generate appropriate predictions for experimental verification. Therefore, the problem of consciousness in robots and artifacts is a recognized problem for neuroscience researchers.

From an AI perspective, since McCarthy (2002), published a seminal paper, many scholars have discussed *machine consciousness*. Similarly, McDermott (2001), devoted his book *Consciousness and Mechanism* to the discussion of *computing theory on consciousness*. Many journals and conference papers discussed the possibility of *consciousness in machines* and robots by proposing theory and architecture. Holland (2004) collected initial attempts at *robot-awareness*. Chella et al. (2020) explored the robot's *self-awareness* through *internal voice*. Scheutz (2014) revisited and discussed the point of contact between *machine consciousness* and *artificial emotion*. A complete study of *self-conscious robots* can be found in Reggia (2013). Among other important works of AI scholars on machine and robot consciousness, the *LIDA architecture* (Franklin et al. 2014) and the *cognitive architecture* proposed by Shanahan (2006), based on the *global awareness workspace model* (Baars 1997). Kuipers (2008) discussed the *conscious experience model* related to learning and sensorimotor interaction in autonomous robots. It is worth noting that Bringsjord et al. (2015), implemented a cognitive system based on *high-order logic* running on a NAO robot that passed the human *self-awareness test*. Beckerle et al. (2018) studied the requirements of *assistive robot technology* for touch. *Robot consciousness* is an important research field, benefiting from many scholars from neuroscience, artificial intelligence, and robotics. As mentioned above, the general feeling is that understanding *biological consciousness* may help build better robots, and research on *robot consciousness* may help understand *biological consciousness* (Fig. 13.1).

13.2 Good Old-Fashioned Artificial Consciousness (GOFAC)

The concept of *robot awareness* has been evolved and aroused great interest due to AI, robot technology, and computer technology development. Chella

Fig. 13.1 Consciousness in robot (Tuchong 2020a)

and Manzotti (2009) discussed many problems in *consciousness* in robot evaluation. These evaluations involve the role of the body, the robot's needs in the environment, its cognitive ability, and the effectiveness of emotional functions. For *robot consciousness*, the most challenging issue is the possibility that the robot may have a *real subjective experience*. However, many of the latest methods in robot consciousness are based on a set of premises that apply research to the so-called *Good Old-Fashioned Artificial Consciousness* (*GOFAC*, Manzotti and Chella 2018).

GOFAC suggests a physical world where *consciousness* appears as a result of a specific *intermediate level*. A theory based on the idea that *consciousness* emerges from an *intermediate level* should explain what that level is and why it produces *consciousness*. However, this interpretation has certain limitations, because the theory does not *transcend consciousness*, but introduces a new level as an intermediate entity, which is obviously less troublesome. On the contrary, the *intermediate level* is an explanatory destruction because it adds two new problems: new level characteristics and consciousness relations. This method can be called *intermediate level fallacy*. It seems appealing because the introduced level looks less threatening and familiar than consciousness itself.

13.3   The Hard Problem of Robot Consciousness

The concept of GOFAC is based on the *hard problems* proposed by philosopher and cognitive scientist Prof. David Chalmers (Chalmers 1996). According to his pioneering work, most research and discussion on *consciousness* is conducted within a conceptual framework that is set by the contrast between *conscious mind* and *cognitive mind*. Such notions are deeply entrenched, subjective, and the gap between the *experience of phenomena* and the *physical attributes*. The tricky issue is: Once all the important facts have been determined, there are still things that need to be explained, which has delayed the understanding of consciousness and put it outside of the robotic realization. If human accept it, the robot will never really realize that, because all physical facts are fixed anyway, there will still be something to add. The acceptance problem is the main reason for the lack of progress in *robot consciousness*.

The premise of *hard problem* (Bilokobylskyi 2019), is that *the subjective and physical attributes are different from each other.* Moreover, this premise is not experimental and may be questioned. In fact, if subjective and physical characteristics are different, it is impossible to face them against each other. For example, consider the comparison between *subjective red* and *true red.* There is no reason to believe that there are two kinds of *red*. Of course, the common saying is that the subjective *red* color is psychological in nature, and in a far-reaching sense, the physical world is inaccessible. He claimed that only by subjective attributes or using equivalent well-known formulas can we experience the amazing characteristics of the phenomena that occur.

The fact that human experience the outside world and their own bodies does not support this claim. Like Chalmers, there is no reason to assume that the perception of the world is different from the physical world. Perception has no subjective attributes, but human experience attributes are composed of the world, and the names we can give to such features are physical. Hard problems are not derived from experience, because if this is true, it cannot be proved by experience. If consciousness is *hard*, it will not affect the physical world. On the contrary, if consciousness is testable, it is not *hard*.

Hard problems are related to *apparent consciousness*, that is, consciousness has no physical role. Accepting *hard problems* means that *consciousness* will be outside of material facts. If *consciousness* is part of the physical world, it can be measured, observed, copied, designed, and implemented in the robot. *Hard problems* motivate people to regard *consciousness* as something hard to solve by scientific means. Consciousness does not have any influence on the physical world, so it is useless from a robotics perspective. However, if consciousness is phenomenal, it will contradict the choice advantages it seems to offer. In addition, all physical events are causally related, which is why they can be measured and observed when they have a causal relationship. There is no such phenomenon in physics and engineering because they are considered unreal automatically. The fact that GOFAC treats *consciousness* as an *apparent phenomenon* is a sign of scientific failure. Once we enter the traditional GOFAC framework, *consciousness* cannot be achieved. The concept of phenomenological awareness seems to be a self-deception hypothesis. Individuals who are the subject of *consciousness* have a feeling that they feel interwoven with the real world. Consciousness is indeed a part of the material world. If there is no consciousness in the current scientific description of the world, then consciousness is incomplete.

Nonetheless, *hard problems* have become famous because they are in stark contrast to the *hard problem* of defining how physical processes cause consciousness. These simple questions are how to explain the ability of a human to recognize faces, produce language, and control behavior. Such disagreements, on the one hand, show that scientists can continue to work without worrying about *consciousness*, on the other hand, *consciousness* is elusive and not bound by the physical world. It also allows engineers, robotics experts, and AI experts to design robotic awareness freely, because they are smart enough to put aside *hard problems* and limit themselves to problems that are easily aware. In GOFAC, *hard problems* create a difference between *hard and weak machine consciousness* (Seth 2009), and it seems possible to focus on functional and ontological problems, respectively. Due to the wide acceptance of the problem, scholars believe that conscious experience is unachievable by science and technology, so they propose a solution. The solution is the illusion that we can focus on the specific problem, that is, these are part of our conceptual framework, and leave the real consciousness problem to some conceptual breakthroughs.

13.4   Major Components of Robot Consciousness

Robot consciousness has so far not succeeded in making progress on phenomenal experience. While the possibility of conscious machines with its ethical implications has repeatedly been addressed, no one has claimed that anything close to a feeling has occurred in an artifact. As before, this persistent and generalized lack of results might be explained by the adoption of familiar and flawed conceptual GOFAC landscape. Here, we will explore five major components of *GOFAC robot consciousness* which include: (1) Functionalism; (2) Information integration; (3) Embodiment; (4) Enaction, and (5) Cognitive mechanisms.

13.4.1 Functionalism

Functionalism is the backbone of *AI consciousness*. The functionalist approach singles out the functional perspective of the mind. Many scholars have elaborated on this criticism in detail, the most famous being philosopher Prof. John Searle (1990) and cognitive scientist Prof. Stevan Harnad (2003). If the mind is a collection of functional relationships, then there is no room for consciousness—the function and feeling of using Hanad's formula. *Functionalism* focuses on the external causality between affairs. Although functionalism is neutral to this causal relationship, it mainly focuses on abstract descriptions that describe reality, which is why it allows multiple implementations. *Functionalism* is a theoretical description of what happens in the system and does not consider the physical components. Therefore, functionalism will never grasp consciousness because it is neutral to the material component of the functional relationship. *Functionalism* will then provide the same description for systems composed of neurons and electronic switches, and the same explanation for conscious and unconscious systems. This is not a fact about consciousness. This is the result of the premise of establishing functionalism.

Functionalism has always been an ideal choice to support the philosophical concept of *zombies*, and zombies are crucial in all the narratives affected by the *hard problem* (Chalmers 1996). A *zombie* is an entity that is indistinguishable from the human on the outside. In terms of its conversation, response, and actions in the world, it is completely unconscious as opposed to a human. The imaginability of zombies tells us more about functionalist limitations than consciousness. There is no evidence that the same physical entity as human may be unconscious. The concept of zombies indicates that the functional description is incomplete, and some key content is missing. In fact, today all machines are considered philosophical zombies. No one expects *Siri* or *Google Assistant* to become zombies.

Many methods of consciousness are functionalist models. Consider the mentioned global workspace model by psychologist Prof. Bernard Baars (Baars 1997) and its implementation (Shanahan 2006; Franklin et al. 2014). The model consists of an appropriate functional structure in which information is concentrated and broadcast. The first step is represented by a specific cognitive structure, the central workspace. The idea of considering a unified concept and a central controller is a neutral concept. The second step is to dilute consciousness, consciousness is just to concentrate on obtaining information.

Another method is the model of consciousness established by cognitive scientist, theoretical and computational psychologist, neuroscientist, mathematician, biomedical engineer, and neuromorphic technologist Prof. Stephen Grossberg (2007, 2017) based on adaptive resonance in the brain. According to the model, due to the feed-forward and feedback connections between the bottom-up and top-down nerves, the state of consciousness in the brain is characterized as a resonant nerve state, that is, neurons excite a mutually synchronized nerve state. In this case, the first step is expressed by proper characteristics of dynamic neural network evolution, that is, the resonance of interconnected neurons is the neutral effect explained by the differential equations that control the dynamics of neural network. The second step is to assert that subjective experience is nothing but this particular state in dynamic neural networks evolution. There is no reason why the centralized access information or resonance state cannot be unconscious.

In summary, this is not to say that robots envisioned by functionalist designers will never be conscious. In fact, no matter what conceptual framework designers use, when they transition from design to implementation, they are constrained by the structure of the physical world. Therefore, its products are not restricted by its conceptual model. Since consciousness is part of the natural manifold, in some cases, regardless of the conceptual framework adopted by its designers, the physical subject structure will succumb to consciousness.

13.4.2   Information Integration

Another popular method in GOFAC is based on the process of finding unique information that generates consciousness. Information at the level of computational processes such as the brain or computer is not part of physical reality. Instead, it is a convenient description. Information is a fictitious entity, such as the center of mass or the meridian. It does not actually exist, but only in our description. It cannot be observed, but it can be calculated a lot.

In terms of information, there is confusion among scientists. The daily inflow of information promotes a general trend to process information as real as water or electricity. However, there is no evidence that information exceeds the physical processes we describe in information terms (Searle 1990). It is just a quantitative description of the causal relationship between events. From a physical point of view, there is no need for an additional level called information above the physical phenomena, but all causal forces are consumed by physical events (Kim 1998; Dowe 2000).

An argument about the fact that information does not physically exist, considering that if the information is true, it should be possible to construct an information detector. Interestingly, it is impossible to construct an information detector. Although the amount of information inside the system can be calculated based on a set of assumptions about how to use the system, it is impossible to detect the amount of information in the system. For example, if we know that a standard CD player will read a CD-ROM, we can calculate its capacity. However, if someone deals with a thing without knowing whether and how to use its physical structure, then there is no way of knowing how much information it contains. This is true in all cases of similar information equipment. Unable to measure mass, charge, length, and other information. Information can be estimated or calculated based on knowledge of a thing and its role in a given context.

Information does not exist unless it is described in some way what happened with the causal event coherent with the original information representation (Shannon 1948). Information is a way to explain causal processes, not a real phenomenon. It is not physical, but causal, undetectable, never measured, only estimated. Besides, there is no law that can explain why a particular state of information should be like a state of consciousness. Information-based awareness methods are still in the middle of inconsistencies. Intermediate entities are now information-sometimes specific information brands like neuroscientist and psychiatrist Prof. Giulio Tononi's *Integrated Information Theory* (Tononi 2004) and its latest version (Oizumi et al. 2014). The effort to downplay is to claim that the attributes of information are those important to consciousness. For example, he claimed that the integrated information is unified, and the consciousness is also unified. There is no description of quality, semantics, content, and all other aspects of our experience. In summary, the method proposed and the idea of generating conscious thought based on information processing are not based on experience, because information does not have physical reality. They are biased against the hope that quantitative, precise methods can provide a scientific framework. These authors emphasized the possibility of measuring consciousness. These methods can estimate successfully the most information states related to consciousness, but so far, it has not been possible to provide reasons for why the information states under review constitute consciousness.

13.4.3 Embodiment

Do we ever think about our body parts: hands, legs and even our whole body are really part of ourselves (embodiment), not an outside agent? or what we called *avatar* nowadays? The truth in the entire concept of *embodiment* can be clearly explained by the renowned *Rubber Hand Illusion (RHI)* experiment.

The *Rubber Hand Illusion (aka RHI)* is a somatosensory illusion reported by Botvinick and Cohen in 1998 (Botvinick and Cohen 1998). It has become almost classic in body awareness research and to study fundamental aspects of bodily sense: *the sense of body ownership* (Moguillansky et al. 2013; Aymerich-Franch et al. 2017; Kalcket and Ehrsson 2012; Slater et al. 2008).

In RHI experiment, a standing screen was positioned beside the arm to hide it from the subject's view with a life-sized rubber model of a left hand and arm was placed on the table directly in front of the subject as shown in Fig. 13.2. The subject sat with eyes fixed on the artificial hand while using two small paintbrushes to stroke the rubber hand and the subject's hidden hand to synchronize brushing timing as closely as possible. After ten minutes, subjects are to complete a two-part questionnaire for an open-ended description of their experience and requested them to affirm or deny the occurrence of nine specific perceptual effects. The completed questionnaires indicated that subjects experienced an illusion which they seemed to feel the touch was not from the hidden brush but that of a viewed brush, as if rubber hand had sensed the touch.

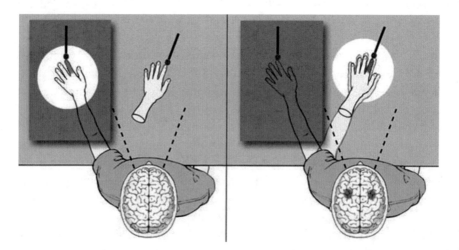

Fig. 13.2 Rubber hand illusion

In robot consciousness, popular methods are related to the notion of embodiment (Holland 2004; Bongard et al. 2006; Shanahan 2006, 2010; Stuart 2010), mainly because they can focus on the robot body. This is a fruitful method that highlights the key features of this embodiment. The body plays a vital role in shaping the interaction between the agent and its environment. Reinforcement of cognition is a mandatory point of view on sensorimotor circuits. However, it is not clear why the implementation should provide clues about how consciousness fits the physical world. The embodiment simplifies inevitably many critical sensorimotor control loops. If the embodiment embodies the fact that the process of cognition or consciousness must be embodied physically, it is a fairly obvious concept. Cognitive processes must be reflected in this sense because any process must correspond to a physical thing and be reflected accordingly. But the meaning of proponents of implementation concepts like Chrisley and Ziemke (2006) is not so trivial. Some researchers compete with traditional cognitive concepts which are high-level processes performed by a central processing unit separate from the body. This method is a branch of historical factors, that is, in most cases, the Cartesian concept of the immaterial mind, the functionalist model of the mind, and the availability of electronic calculators before they are combined with artificial objects. All these factors motivate people to have a specific concept of psychology and its processes. However, whether in the philosophical debate or in the technical field, they are no longer relevant.

AI is biased by the Cartesian view of thinking. The implementation allows AI scholars to emphasize the physical properties of agent headscarves. However, this fact does not mean that the body is the only component of the medium. The concept of implementation contradicts itself. In fact, this embodiment was considered to get rid of non-physical thoughts, because the body and its interaction with the world seem to be a feasible solution. Unfortunately, the concept of the *body* is unclear. Generally, an object is a subject only when it is a subject. However, the body concept is the cornerstone of the subject on the circumference. The body is another intermediate entity that should bridge the gap between the world and consciousness. This is a symptom of inconsistency at the intermediate level. It is recommended to use the body or its interaction with the environment as an intermediate level. At the same time, the fade step processes the body as if it were more than just moving physical objects. Of course, the last step is the relevance of robot consciousness because the researchers do not have organisms. There is no description of a qualified object characteristics. In this sense, the washing machine can be viewed as a whole because it responds to external stimuli, swallowing, processing, discharging, energy consumption, and planning. Anthropomorphic robots have the same view (Netherlands, 2004). In summary, the embodiment attempts to take advantage of the inconsistency of the middle layer by adopting the ambiguity concept of the body and reduce consciousness to something more ordinary.

13.4.4 Enaction

Enaction can provide another feasible solution to realize robot consciousness because it shows that experience is composed of the body, its interaction, and its relationship with the world so it can be realized in artifacts (O'Regan and Nöe 2001). Activism defends a firm position that together with implementation it may produce results in many fields especially in cognitive science (Stewart et al. 2010).

Consider the basic tenet of enaction in philosopher Prof. Alva Noë's formulation:

> Perceiving is a way of acting [...] What we perceive is determined by what we are ready to do [...] We enact out perception; we act it out (Noë 2004, p. 1).

Noë suggested in his work that the intermediate level based on movement should deepen people's understanding, but it cannot naturally explain why the movements performed by human body should be different from those performed by robots or animals. Unless by referring to the subject, inactivism cannot provide a standard to distinguish between real and simple actions. In other words, behavior refers to actions performed by the subject with intention and understanding. Then, there is a specific cyclical risk in their argument.

Consider this point in philosopher Prof. John Stewart's formulation:

> How can a material state be a mental state? Hoary it may be, yet the problem is anything but solved. [...] The paradigm of enaction solves this problem by grounding all cognition as an essential feature of living organism (Stewart 2010).

Of course, as he admitted, this does not solve the problem. It only shifts the burden of explaining the concept of organisms. Since vitalism has long been abandoned, the emphasis on life and biology does not seem to have a convincing conceptual fulcrum. In this way, the suggested middle layer is the organism, whose feedback circulates with the outside world. Why these phenomena are so special remains to be explained. This is the second step of inconsistency.

Again, to reach the intermediate level of understanding, namely understanding and sensorimotor knowledge. In the physical world, what is this intermediate level and why the effect of exercise on sensory stimulation should lead to conscious experience is unclear.

13.4.5 Cognitive Mechanisms

The most obvious candidates are awareness and intelligence. Here, we have a promising intermediate entity that does not seem to be demanding, and we may consider whether it is appropriate. After all, there seems to be a close connection between *cognitive ability* and *consciousness*. In most cases, when human impose higher-order cognitive processes, they are conscious. However, it is fair to assume that in many cases when a person is conscious, very weak intelligence is required, or that many of the most creative ideas are the result of most unconscious activities (Lavazza and Manzotti 2013).

The fact is that many scholars tend to focus on intelligence and cognition, and expect that once all practical problems are solved, consciousness will be free. Instead, many people think that the problem will disappear as a wrong one. Moreover, cognition is an intermediate level which may lead to conscious knowledge not conscious experience in this case. This is a symptom of inconsistency at the intermediate level.

13.5 The Intermediate Level of Robot Consciousness

Scholars working in robot consciousness suggest an *intermediate level*— sensory-motor patterns, information, cognition, global workspace—as a possible explanation for consciousness. What is missing is why such a level should lead to consciousness. From an epistemic perspective, it is as though they suggested an explanans without providing its relationship with the explanandum, i.e. consciousness. Table 13.1 summarizes different GOFAC landscapes on robot consciousness.

The hard problem, GOFAC's method and the *strong versus weak AI* consciousness argument are grouped by a common factor, because they all deal with the lack of any causal consciousness in the world. Considering the problem that leads to the zombie problem, agents with cognitive equivalence lack awareness. In turn, GOFAC will not solve the subjective experience

Table 13.1 Different levels of GOFAC landscapes on robot consciousness (Manzotti and Chella 2018)

	Different levels GOFAC on robot consciousness		
	Physical world	Intermediate level	Watered down version
Functionalism	The physical states that realize functional structures	Global workspace, centralized representations, adaptive resonance	Access consciousness
Information and computation	The physical states that transmit causal processes	Integrated Information	Integration consciousness
Embodiment	Objects	Body states, body-world states	Sensory-motor loops
Enaction	Interactions between objects and environment	Actions	Knowledge of sensory-motor loops
Cognition	Brain or processor	Cognitive states	Knowledge

problem. Finally, the gap between the machine consciousness of weak and strong is considered to be used to deal with the cognitive process without solving the crux of the problem which is the conscious experience. Weak consciousness aims to deal with the problem of consciousness function. Those who have a causal relationship have no strong consciousness.

New assumptions about the nature of the physical world are needed. Consciousness is the fact that we need to find our place in nature. So, if consciousness is neither a previously checked process, what is left? The suggestion is that consciousness is the structure of the physical world itself. Like Perlovsky (2016), this move was excluded. The concept of the physical world must be fundamentally wrong. The possible error may be the location of something called consciousness, not the location of the subject or neuron/computing structure. Another error may be contained in the separation between subject and object. This article shows that GOFAC will never achieve machine consciousness, therefore, it strongly requires a strong conceptual framework to replace the puzzle and its cognition.

Naturally, when the goal is to design a conscious robot it is necessary to look for consciousness in the physical world. Robots have no resources other than those provided by the physical world. It sounds cliché but all the mentioned methods operate on this principle. Therefore, any feasible solution needs to set aside the premise that has prevented any progress so far, that is, the general idea that consciousness is different from the physical world. This issue needs to be reconsidered from the beginning.

We believe that it is possible to enrich a fundamental alternative that will set aside the annoying theoretical framework arising from the adoption of the *hard problem*. First, we make consciousness the same as all other physical properties around us, which can be measured and observed. In addition, consciousness is causally active and is in time and space. Finally, it is composed of matter or energy. These premises are nothing more than reiterating the assumption that consciousness is the body. In fact, in physical space–time positioning, everything related to cause and effect is composed of matter/energy and is observable.

As far as most scholars are concerned, if they realize that they are invisible in the physical world, then this move will naturally be considered infeasible. Neuroscientists have been searching for it in the brain for centuries but have not found any rebalancing. There is nothing like a conscious experience in the brain, so there is no inside the machine. The solution may require a conceptual leap.

Consider the possibility that although consciousness is physical, it is not actually in the agent, whether it is biological or artificial. The suggestion is that awareness is the same as the external object handled by the agent. In this way, the physical attributes of the external world may be the same as the attributes of conscious experience (Manzotti 2006, 2017; Manzotti and Chella 2018).

An example will help. The agent, that is, the body of a creature or robot, is interacting with an external object e.g. a yellow banana. There is no banana inside the potion, it is yellow, elongated and slightly curved. When we look for the agent's experience in the agent, we will be forced to conclude that the characteristics, i.e. yellow, slender, curved in the agent do not have any physical properties. Unable to detect something like the experience in our body, we may conclude that consciousness is indeed intangible, phenomenal, and cannot be measured directly in an extraordinary world. This option is used for *hard problems* and all related methods.

We suggest an alternative method. When the agent interacts with the banana, according to the agent's experience, there is an appropriate physical entity, namely the banana itself. Like its experience, it is yellow, slender and slightly curved. The banana experience is all-encompassing. It is better than anything we might want to find in the agent. External objects score higher than any internal representation.

In terms of machine consciousness this method has many advantages. The debate has neither biological materials nor emerging property. There are no so-called suspicious attributes that cannot be observed physically. Everything is measurable, observable, and crucially relevant, not additive.

13.6   Case Study: Are We Self-consciousness Robots?

In this chapter, we learnt GOFAC five major components in robot consciousness: global workshop, integrated information, embodiment, enaction, and cognitive mechanisms. Naturally, we as human believe that we are self-consciousness, self-aware beings can 100% fulfill these requirements. Is it possible that we not only live in illusion or projection per Plato's famous *cave allegory*, but also as robot agents of a global chessboard?

Separate into two groups of students and debate on the topic: Are we self-consciousness robots?

Here are certain points for considerations (Fig. 13.3):

1. List out GOFAC five major requirements. For each of them, compare between us and robots against these basic requirements.
2. Is it possible that we live in an artificial global workspace with shared consciousness?

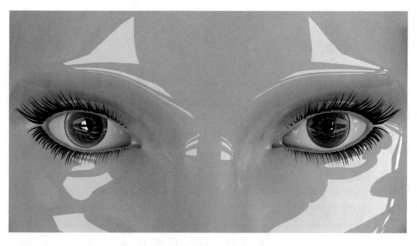

Fig. 13.3   Are we self-consciousness robot? (Tuchong 2020b)

3. Revisit *Plato's mind–body dualism*, is it possible that the so-called mind–body association nothing more than the embodiment experience alike RHI (Rubber Hand Illusion) studied in this chapter?
4. When we consider cognitive mechanism, can it be simulated artificially? If yes, what does it mean?

13.7 Conclusion

In this chapter, we study one of AI most challenging topics—the possibility of self-consciousness and self-aware robots (or AI systems). Since Turing Test proposed by Sir Alan Turing in 1950s, to the latest AI and robotics development, we have never ceased to explore human intelligence and the creation of intelligent systems and robotics. With the advancement of technology, almost all core AI functionalities include machine learning, computer vision, data mining, natural language processing, and agent ontology have been examined immensely with significant success and applied to our daily activities. Many AI scientists believe it is the time to explore this ultimate question—robot consciousness.

The intermediate level of GOFAC five major components: global workspace, integration of information, enaction, embodiment and cognitive mechanisms were studied extensively in the past decades with certain degrees of contribution. There is a fine chance self-aware and self-consciousness robots will appear and live among us in the near future. If it is really happened, says in 20 years, it is necessary for us to consider another critical AI and robotics issue—AI ethics, which will be discussed in the next chapter. A final remark: If we can create a *"truly"* self-consciousness and self-aware robot which is in a certain degree per human, how do we know we are not conscious robots created by our creator?

References

Aymerich-Franch, L., et al. (2017). Non-human looking robot arms induce illusion of embodiment. *International Journal of Social Robotics, 9*(4), 479–490.

Baars, B. J. (1997). *In the theater of consciousness. The workspace of the mind*. Oxford: Oxford University Press.

Botvinick, M., & Cohen, J. (1998). Rubber hands 'feel' touch that eyes see. *Nature, 391*(6669), 756–756. https://doi.org/10.1038/35784.

Beckerle, P., et al. (2018). Feel-good robotics: Requirements on touch for embodiment in assistive robotics. *Frontiers in Neurorobotics, 12,* 84. https://doi.org/10.3389/fnbot.2018.00084.

Bilokobylskyi, O. (2019). "The hard problem" of consciousness in the light of phenomenology of artificial intelligence. *Skhid, 1*(159), 25–28.

Bongard, J., et al. (2006). Resilient machines through continuous self-modeling. *Science, 314,* 1118–1121.

Bringsjord, S., et al. (2015). Real robots that pass human tests of self-consciousness. In *24th IEEE International Symposium on Robot and Human Interactive Communication (RO-MAN)* (Kobe) (pp. 498–504).

Chalmers, D. J. (1996). *The conscious mind: In search of a fundamental theory.* New York, NY: Oxford University Press.

Chatila, R., et al. (2018). Toward self-aware robots. *Frontiers Robotics AI, 5,* 88. https://doi.org/10.3389/frobt.2018.00088.

Chella, A., & Manzotti, R. (2009). Machine consciousness: A manifesto for robotics. *International Journal of Machine Consciousness, 1,* 33–51.

Chella, A., et al. (2020). Developing self-awareness in robots via inner speech. *Frontiers in Robotics and AI, 7.* https://doi.org/10.3389/frobt.2020.00016.

Chrisley, R., & Ziemke, T. (2006). *Embodiment in: Encyclopedia of cognitive science.* Hoboken, NJ: Wiley, Ltd.

Cosmin, C. T. (2017). The conscious life—The dream we live in. *Dialogo, 3*(2), 65–71. https://doi.org/10.18638/dialogo.2017.3.2.5.

Cottingham, J. (2017). *Descartes: Meditations on first philosophy: With selections from the objections and replies* (Cambridge Texts in the History of Philosophy). Cambridge University Press.

Damasio, A. (2010). *Self comes to mind: Constructing the conscious brain.* New York, NY: Pantheon Books.

Dehaene, S. (2014). *Consciousness and the BRAIN: Deciphering How the brain codes our thoughts.* London: Penguin Books.

Dowe, P. (2000). *Physical causation.* New York, NY: Cambridge University Press.

Franklin, S., et al. (2014). LIDA: A systems-level architecture for cognition, emotion, and learning. *IEEE Transactions on Autonomous Mental Development, 6,* 19–41.

Grossberg, S. (2007). Consciousness CLEARS the mind. *Neural Networks, 20,* 1040–1053.

Grossberg, S. (2017). Towards solving the hard problem of consciousness: The varieties of brain resonances and the conscious experiences that they support. *Neural Networks, 87,* 38–95.

Harnad, S. (2003). Can a machine be conscious? How? *Journal of Consciousness Studies, 10,* 67–75.

Hildt, E. (2019). Artificial intelligence: Does consciousness matter? *Frontiers in Psychology, 10,* 1535. https://doi.org/10.3389/fpsyg.2019.01535

Holland, O. (2004). The future of embodied artificial intelligence: Machine consciousness? In F. Iida (Ed.), *Embodied artificial intelligence* (pp. 37–53). Berlin: Springer.

Kalckert, A., & Ehrsson, H. (2012). Moving a rubber hand that feels like your own: A dissociation of ownership and agency. *Frontiers in Human Neuroscience, 6,* 40. https://doi.org/10.3389/fnhum.2012.00040.

Kim, J. (1998). *Mind in a physical world*. Cambridge, MA: MIT Press.

Kuipers, B. (2008). Drinking from the firehose of experience. *Artificial Intelligence in Medicine, 44,* 55–70. https://doi.org/10.1016/j.artmed.2008.07.010.

Lavazza, A., & Manzotti, R. (2013). An externalist approach to creativity: Discovery versus recombination. *Mind & Society, 12,* 61–72.

Legg, S., & Hutter, M. (2007). A collection of definitions of intelligence. In *Proceedings of the 2007 Conference on Advances in Artificial General Intelligence: Concepts, Architectures and Algorithms* (pp. 17–24). Amsterdam: IOS Press.

Manzotti, R. (2006). A process oriented view of conscious perception. *Journal of Consciousness Studies, 13,* 7–41.

Manzotti, R. (2017). *Consciousness and object. A mind-object identity physicalist theory*. Amsterdam: John Benjamins Publishing Company.

Manzotti, R., & Chella, A. (2018). Good old-fashioned artificial consciousness and the intermediate level fallacy. *Frontiers Robotics AI, 5.* https://doi.org/10.3389/frobt.2018.00039. https://www.frontiersin.org/articles/10.3389/frobt.2018.00039/full.

McCarthy, J. (2002). *Making robots conscious of their mental states*. Retrieved May 29, 2020, from https://jmc.stanford.edu/articles/consciousness/consciousness.pdf.

McDermott, D. (2001). *Mind and mechanisms*. Cambridge, MA: MIT Press; Bradford Books.

Moguillansky, C. V., et al. (2013). Exploring the subjective experience of the "rubber hand" illusion. *Frontiers in Human Neuroscience, 7*(659), 659.

Noë, A. (2004). *Action in perception*. Cambridge, MA: The MIT Press.

Oizumi, M., et al. (2014). From the phenomenology to the mechanisms of consciousness: Integrated information theory 3.0. *PLoS Computational Biology, 10,* e1003588.

O'Regan, K., & Noë, A. (2001). A sensorimotor account of visual perception and consciousness. *Behavioral and Brain Sciences, 24,* 939–1011.

Pandey, S. C. (2018). Can artificially intelligent agents really be conscious? *Sādhanā, 43*(7), 1–17.

Perlovsky, L. I. (2016). Physics of the mind. *Frontiers in Systems Neuroscience, 10,* 84. https://doi.org/10.3389/fnsys.2016.00084.

Raoult, A., & Yampolskiy, R. (2015). *Reviewing tests for machine consciousness*. Retrieved May 29, 2020, from https://www.researchgate.net/publication/284859013_DRAFT_Reviewing_Tests_for_Machine_Consciousness.

Reggia, J. A. (2013). The rise of machine consciousness: Studying consciousness with computational models. *Neural Networks, 44,* 112–131.

Scheutz, M. (2014). Artificial emotions and machine consciousness. In K. Frankish & W. Ramsey (Eds.), *The Cambridge handbook of artificial intelligence* (pp. 247–266). Cambridge, MA: Cambridge University Press.

Seth, A. K. (2009). The strength of weak artificial consciousness. *International Journal of Machine Consciousness, 1,* 71–82.

Shanahan, M. (2010). *Embodiment and the inner life: Cognition and Consciousness in the space of possible minds.* Oxford: Oxford University Press.

Shanahan, M. P. (2006). A cognitive architecture that combines internal simulation with a global workspace. *Consciousness and Cognition, 15,* 433–449.

Shannon, C. E. (1948). A mathematical theory of communication. *Bell System Technical Journal, 27,* 623–656.

Searle, J. R. (1990). Is the brain a digital computer? *Proceedings of the American Philosophical Society, 64,* 21–37.

Slater, M., et al. (2008). Towards a digital body: The virtual arm illusion. *Frontiers in Human Neuroscience, 2,* 6. https://doi.org/10.3389/neuro.09.006.2008.

Stewart, J. (2010). Foundational issues in enaction as a paradigm for cognitive science: From the origin of life to consciousness and writing. In J. Stewart, O. Gapenne, & E. Di Paolo (Eds.), *Enaction. Toward a new paradigm for cognitive science* (pp. 1–31). Cambridge, MA: The MIT Press.

Stewart, J., et al. (2010). *Enaction.* Cambridge, MA: The MIT Press.

Stuart, S. A. J. (2010). Conscious machines: Memory, melody and muscular imagination. *Phenomenology and the Cognitive Sciences, 9*(1), 37–51.

Tononi, G. (2004). An information integration theory of consciousness. *BMC Neuroscience, 5,* 42.

Tononi, G. (2012). *Phi: A voyage from the brain to the soul.* New York, NY: Pantheon Books.

Tuchong. (2020a). *Conscious in robot.* Retrieved June 1, 2020, from https://stock. tuchong.com/image?imageId=903133155985457231.

Tuchong. (2020b). *Are we self-consciousness robot?* Retrieved June 1, 2020, from https://stock.tuchong.com/image?imageId=467194688471695854.

14

AI Ethics, Security and Privacy

On the path to ubiquity of AI, there will be many ethics-related decisions that we, as AI leaders, need to make. We have a responsibility to drive those decisions, not only because it is the right thing to do for society but because it is the smart business decision.
Dr. Rana el Kaliouby (AI Scientist, born 1978)

Abstract AI and robot technology of the new era will have a major impact on human development in future. Such technologies are beginning to raise some basic issues about how to use these systems and what are potential risks and how to control them. This chapter explores one of AI and robotics popular and controversial topic—AI ethics. First, we introduce AI ethics with Asimov's Three Laws of Robotics. Next, we study major aspects and concerns related to AI ethics include: robots ethics, robot rights, moral agents, opaqueness of AI systems, privacy & AI monitoring, automation and employment, prejudices in AI systems, responsibility for autonomous machines, and international AI ethic policy.

AI and robot technology are technologies of the new era and will have a major impact on human development in future. Such technologies are beginning to raise some basic issues about how to use these systems and what are potential risks and how to control them. In this chapter, we will explore this important theme—AI ethics. First, we will introduce an overview of AI ethics background through *Asimov's Three Laws of Robotics*. Then, we will study

R. S. T. Lee, *Artificial Intelligence in Daily Life*,
https://doi.org/10.1007/978-981-15-7695-9_14

the main aspects and concerns related to AI ethics, including robot ethics; robot rights; moral agents; opaqueness of AI systems; privacy and AI monitoring; automation and employment; prejudices of AI systems; autonomous machines responsibilities; and international AI ethics policies.

14.1   An Introduction to AI Ethics

We explored various AI technologies such as machine learning, data mining, computer vision, NLP, and agent ontology in the first 13 chapters. In theory, all these AI-enabled technologies are sufficient to build intelligent systems, from self-driving vehicles to domestic robots in smart houses. So, what do we need to consider next is how to control this technology? Imagine that one day, when we used an autonomous vehicle to work and caused an accident. What are the responsibilities? Is it the driver (intelligent machine)? Owner (human)? Or the smart car manufacturer? In more serious cases, what happens if the surgical robot causes an accident and the patient dies during the operation. What is the responsibility?

Artificial intelligence and machine learning technologies are changing society rapidly. It is almost certain that they will continue to do so for decades to come. These powerful new technologies will promote social change and moral impact. Artificial intelligence provides us with everything that human already have whether virtuous or vicious. Many things are at stake. We should think very carefully about how to make this transition.

In this chapter, we will study various important aspects related to AI ethics, including: robot ethics; robot rights; moral agents; opaqueness of AI systems, privacy and AI monitoring, automation and employment, prejudices of AI systems, autonomous machines responsibilities, and international AI ethics policies (Müller 2020) (Fig. 14.1).

14.2   Asimov's Laws of Robot Ethics

Anyone familiar with AI and robotics might have heard the famous *Three Laws of Robot* in remarkable works *Runaround*—a science fiction written by Prof. Issac Asimov (1920–1992) in 1942 (Asimov 1942; Anderson 2008):

1. A robot may not injure a human being or, through inaction, allow a human being to come to harm.

Fig. 14.1 AI ethics (Tuchong 2020a)

2. A robot must obey the orders given it by human beings except where such orders would conflict with the First Law.
3. A robot must protect its own existence as long as such protection does not conflict with the First or Second Laws.

Robot ethics also known as *roboethics* involve ethical issues that occur with robots or AI systems such as the long-term or short-term threats of robots to human, and whether there are problems with the use of certain robots such as *healthcare robots*, and how to design the robot to make it *ethical*.

In addition, as robots become more advanced, robot ethics also refer to human behaviors for robots (Veruggio and Fiorella 2008). Robot ethics is a subfield of AI technology. Researchers from various fields have begun to solve ethical issues related to ways to create robotic technologies and implement them in society to ensure human safety (Müller 2020).

The British Standards Institute (BSI) provides an excellent article specially prepared for robot creators to ensure that their intelligent machines are *ethical*. Document BS8611 (BS6110 2016) describe robot ethics as: *Robots should not be designed solely or primarily to kill or harm humans; humans, not robots, are the responsible agents; it should be possible to find out who is responsible for any robot and its behavior.* The regulations include the following principles: All robotics must not contain cultural, gender or status discrimination. The document questioned whether robots should be designed to promote emotional bonds between users and warned rogue machines for the possibility of changing their own code.

Most of the worries that prompted the need for these rules came from the inspiration of philosophers and writers such as Asimov and his works. As robots have become more autonomous, AI has surpassed human capabilities in many ways, and the need for robot ethical standards has become more urgent (Lin et al. 2011). Futurists and technical experts, such as Mr. Elon Musk, Mr. Steve Wozniak, and Prof. Steven Hawking (1942–2018) have expressed concern that if not controlled, robots may one day lead to the collapse of humanity. More optimistic views include the hope that well-designed robots can help the world recover from human problems (Turner 2018).

14.3 Robot Rights

We studied robotics and intelligent agents' *self-awareness* and *self-consciousness* in Chap. 13. These perceptual and self-conscious autonomous robots may live and work with us in future. Technically, they should enjoy *human rights* just like us for moral reasons.

Robot rights are the notions that humans ought to take moral responsibility for intelligent robots they created. It has been advocated that the rights of robots including to the right to survive and exercise their own faults can be related to human rights enjoyed by humans in society and robots to serve human obligation. These may include the right to life and freedom, the freedom of thought and expression, and equality of all beings before the law. Many AI scientists disagree whether specific or detailed laws are required or implemented as quickly as possible. Glenn McGee (2007) reported that there will be enough humanoid robots by 2020 and Ray Kurzweil set the date to 2029 in his book: *Singularity is Coming* (Kurzweil 2007). The rules of 2003 Lebner Prize Contest (Prestigious AI Contest) envisage the possibility of robots having their own rights as stipulated in their regulations (Article 61):

> If, in any given year, a publicly available open source Entry entered by the University of Surrey or the Cambridge Center wins the Silver Medal or the Gold Medal, then the Medal and the Cash Award will be awarded to the body responsible for the development of that Entry. If no such body can be identified, or if there is disagreement among two or more claimants, the Medal and the Cash Award will be held in trust until such time as the Entry may legally possess, either in the United States of America or in the venue of the contest, the Cash Award and Gold Medal in its own right. (Loebner Prize competition regulation 2003).

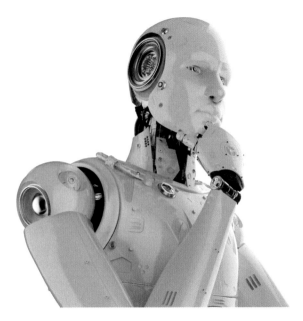

Fig. 14.2   Robot rights (Tuchong 2020b)

One typical example is *Android Sophia* granted the title of Honorary Citizen of Saudi Arabia in 2017. Although some observers considered it as a public relation show more than a serious lawful recognition, it created a new chapter in *robot rights*. After years of moral disputes over robot rights and ethics, sentientism philosophy, in fact, gives all sentient robots the degree of moral consideration, primarily human and most nonhuman species. If AI or even extraterrestrial intelligence shows perceptual evidence, the philosophy believes that they should be treated equally with respect and legal rights (Müller 2020) (Fig. 14.2).

14.4   Moral Agents

If *roboethics* are concerned with moral issue, these robots (or agents) are known as *moral agents* (Brożek and Janik 2019; Cervantes et al. 2020). Some AI scientists have used *moral agents* in a lower sense, due to the use of intelligent agents in software engineering with no responsibilities and rights will appear (Allen et al. 2005). Prof. James Moor (2006) distinguished four types of machine agents: (1) moral influence agents, (2) implicit moral agents, (3) explicit moral agents, and (4) complete moral agent. He explained that explicit moral agents can make clear moral judgments and have the ability to justify them reasonably as ordinary adults. Using this term, several methods have been proposed to achieve moral agents through the use of *programming methods* (so-called *operational morality*) or self-development ethics (so-called *functional morality*) which aim to achieve complete morality through intelligence and sentiment development.

The development of robotics and autonomous agents raises a critical question: Are artificial systems morally responsible? It involves two aspects (Müller 2020):

(1) *AI and robotics cannot be morally responsible*

Prof. Batya Friedman et al. (1992) believes that *intention* is an essential factor for ethical responsibility, and AI systems developed in 1992 cannot be intentional. Prof. Arthur Kuflik (1999) pointed out that we must be accountable for ethical responsibility for any decision done by the intelligent systems, because we are the originators who design these systems. He further maintained that human should never give up computer supervision. Prof. Patrick Hew (2014) pointed out that for AI systems to bear ethical responsibility, their behaviors must not be entirely controlled by an outsider, otherwise, such AI systems should not bear any responsibility for their actions. Ethical responsibilities are assigned to those who create and program the system.

(2) *AI and robotics can be morally responsible*

Distinguished Prof. Colin Allen et al. (2005) pointed out that if the behavior of a man-made system is indistinguishable from the function of a *moral person*, then it may be morally responsible, which raises the idea of *Moral Turing Test*. Prof. Andreas Matthias (2004) describes the responsibility gap under which it is unjust to hold people accountable for machines but holding machines accountable will challenge traditional attribution. He proposed three situations in which the behavior of the machine should be attributed to the machine, not to the designer or operator of the machine: (1) Modern machines are inherently unpredictable; (2) As manual coding procedures are replaced by more sophisticated means, there is more and more *fuzziness* between manufacturers and systems; (3) There are intelligent systems in which their operating rules can be self-altered during operation.

14.5   Opaqueness of AI Systems

Opaqueness and prejudices are the core issues that are now sometimes referred to as *Data Ethics* in the research of big data and data mining (Müller 2020; Xafis and Labude 2019). AI systems for automatic decision-making and predictive analysis trigger issues on accountability, auditing, and due process. Besides, it is usually difficult for individuals being involved to tell how the system can come up with such decisions. In other words, AI systems are totally obscure to them. If such AI systems apply data mining technology, it is even obscure to AI experts to know what patterns actually are. This opaqueness exacerbates prejudice in decision-making systems and datasets. Therefore, at least in the hope of eliminating prejudice, opacity analysis and prejudice with the political response to solve these two problems.

AI systems usually make use of data mining technique such as deep networks to extract patterns from problem domain whether or not they provide the correct solution in reality. These intelligent systems usually capture patterns in the data and mark them in certain ways that are useful to the decisions of these technologies, and in most situations, even AI scientists do not really know which data patterns are used by the system. The program is constantly progressing with new input data, the mode used by the machine learning system will be altered accordingly. In other words, the results will be obscure to the end-user or even AI scientists themselves. In addition, the quality of the program is highly affected by input data. If the data being involved are biased, the program will be very likely to reproduce such bias as well. There have also been a large number of literature on machine learning systems limitations recently which are essentially complex data filters (Marcus 2018). Some individuals even believe that today's AI ethical issues are technological shortcuts taken by AI itself (Cristianini 2020).

14.6   Privacy and AI Monitoring

A general discussion on privacy and AI monitoring (Macnish 2017) mainly involving admittance to privacy data and information. There are several recognized aspects of the right to privacy, such as the *right to own*, not to mention the *right to information privacy*, *the right to privacy as part of the personality*, and *the right to control and confidentiality of their own information* (Bennett and Raab 2006). Privacy research has focused on AI monitoring of secret services to the country traditionally, but now it includes AI monitoring of personnel, companies and even individuals from other countries. AI technology has undergone major changes and regulations have responded slowly over the past few decades. As a result, some anarchy has been exploited by the most powerful participants sometimes even knowing everything (Müller 2020).

In the digital world, dictating who have the right to acquire privacy information and who can make use of such information is much more difficult to determine than the old days especially for today's AI systems. For instance, the facial recognition system by using photo-images or video-shots can identify personal data and search for individuals anytime and anyplace without our notice nowadays. It extends to other recognition technologies such as device fingerprint recognition which is very common on the Internet and is sometimes shown in privacy policies. How the data trail we left paid for free services but were not informed of the value this privacy information when we were asked to release such information? Part of the company's privacy information acquisition seems to be based on deception, using people's weaknesses to exacerbate procrastination, addiction, and manipulation. In this AI monitoring economy, the main objective of these social networks, media, and internet platforms are to acquire, maintain our privacy information and data. AI monitoring now becomes the usual practice of these social networks and internet/mobile platforms. Citizens today seem to have lost the degree of autonomy they need to perform daily activities. If ownership is the correct relationship here, we will lose data ownership. Figure 14.3 shows the scene of AI monitoring system in a smart city. Privacy protection technologies can largely obscure the identity of individuals or groups which has become the main standard of data science. They include anonymization and access control with encryption. Although these technologies require cost and more effort, they can avoid many privacy issues. One of the main practical

Fig. 14.3 AI monitoring and surveillance (Tuchong 2020c)

difficulties is to implement laws and regulations at the state and individual levels.

14.7 Automation and Employment

Obviously, AI and robotics will bring huge growth in overall productivity and wealth. Attempting to increase productivity is usually an economic feature, and the emphasis on *growth* is a universal phenomenon in modern world (Harari 2016). However, the productivity gains achieved through automation usually mean that the same output requires less manpower. This does not necessarily mean unemployment because the increase in available wealth can fully drive demand to offset the increase in productivity. Classical automation replaces human muscles while software-automation resulting with human decision-making and judgement process replacement which cause a fundamental change in the labor market. The main issues are: Will the impact be different this time? Will employment and wealth development keep up with employment elimination? What is the transition cost even if it is insignificant?

At present, due to AI results and robotics automation, the *polarization of work* appears in the labor market (Goos et al. 2009): *The demand for high-skilled technical jobs is high, high-paying, low-skilled job demand is low and salaries are not high, but moderately qualified jobs have a very good chance to be replaced by automation.*

Generally speaking, the unemployment issue is a question of how a society should allocate commodities reasonably. Rawls (1971) believes that the principle of choice will support basic freedom and distribution, which is the greatest benefit to the most vulnerable members of society. In fact, AI-based economy has three main features that cause this kind of justice impossible: (1) Such economy is functioned in an environment that is difficult to regulate; (2) Such economy is functioned in a market with winners and the monopolist will develop all advantages quickly; (3) Such economy is mainly based on intangible-assets (Haskel and Westlake 2017). In other words, it is technically impossible to control these multinational IT companies as they are not bounded by any specific location. These features in a certain degree imply that if we allow wealth distribution regulated by the market itself. It might cause unfair resources distribution as such a scenario has already happened in many countries nowadays (Fig. 14.4).

Fig. 14.4 Automation and employment (Tuchong 2020d)

14.8   Prejudices of AI Systems

An AI-based *decision-making* and *predictive-analysis* system run on input data and generate decisions. The output can range from relatively trivial to very important results. For example, this restaurant meets our preference, or the loan application has been rejected. *Predictive Analysis* is usually applied in business, insurance, medical and other related fields to predict future development. Predictive policing uses predictions and many people worry that this may cause many issues in *bias and prejudices* (Ferguson 2017) as it might violate an individual's freedom for which their future behavior is wrongly predicted. For example, a future AI predictive system might predict someone has a very high chance to commit a crime. The law enforcement officers, as a result, will use such predictive system to forecast and even punish actions instead of waiting for crimes to occur alike the sci-fi movie *Minority Report* in 2002. The core issue is that these predictive AI systems might cause deviations that already exist in data used to build the system or are biased by some wrong initial assumption. So, whether this is a problem rely on the *level of trust* in the technical quality of these predictive systems and on objectives assessment by such systems as well (Fig. 14.5).

When making unfair judgments, *prejudices* usually surface because individual judgments are usually affected by characteristics that are not related to things in front of you and are usually discriminatory prejudices against group members. Therefore, one form of prejudice is a person's learning cognitive characteristics usually not clearly stated. The people involved may not be aware of this prejudice, they may even object to their being found to be prejudiced honestly and explicitly. Another form of deviation in data is when it reveals methodical errors. Technically speaking, any input dataset might only

Fig. 14.5 AI and Predictive Analytics (Tuchong 2020e)

address a single type of problem; so creating a dataset only involves the risk of using it for other types of problems and then prove that the dataset is biased. Not only does machine learning fail to identify prejudice, but it can also organize and automate *historical prejudice.* In order to detect and eliminate prejudices in AI systems, we have to make tremendous technical efforts, but it is still in the early stages and a lot of effort is needed to avoid such bias and prejudices (Müller 2020).

14.9 Responsibility for Autonomous Machines

One major concern in AI is the degree of *autonomous robots'* improvement. Is this a problem that our current conceptual solution should apply, or it only needs certain system adjustments? Most authorities have sound civil and criminal liability systems to solve such problems. Technical standards such as machine safely in the medical environment might need to be regulated. International organizations such as BSI (British Standards Institution) and IEEE (Institute of Electrical and Electronics Engineers) have developed related standards on security and related technical standards for autonomous machines. Let us take a look on autonomous driving as an example.

Self-driving vehicles such as smart cars are expected to reduce manmade driving injuries, including fatalities in traffic accidents and injuries because of human errors. However, there are questions about how selfdriving cars should behave and their responsibilities and risks should be disseminated among their complex systems. Ethical driving issues such as speeding, dangerous overtaking, unsafe distances, etc. are classic issues in pursuit of personal and common interests. These driving issues have been

Fig. 14.6 Autonomous smart car in smart city (Tuchong 2020f)

widely covered in laws and regulations. However, programming the car so that it follows the rules rather than passengers' interests or achieves maximum utility is forced to solve the standard programming ethical machine problems (Fig. 14.6).

14.10 International AI Ethic Policy

Although public discussion about AI ethics is fierce, the actual technology policy is hard to devise and implement. It can be achieved in many ways such as infrastructure and funding to various regulatory agencies and laws. Artificial intelligence policies may conflict with other technologies or general policy directives.

Although actual policies rarely occur, there are some good starts: the new EU policy document recommends that *Trusted-AI* should be: legal, ethical, and technically robust and then clarify it as 7 major requirements: (1) human supervision; (2) system robustness; (3) data privacy; (4) system transparency; (5) fairness; (6) happiness; and (7) responsibility (AI HLEG 2019). Since the advent of nuclear power, more European research has now been conducted under *Responsible Research and Innovation (RRI)*. Professional ethics is also a standard area of information technology including issues covered in this chapter.

Fig. 14.7 AI Codes of Ethics (Tuchong 2020g)

14.11 Case Study: Do We Need a New Set of AI Ethic Rules?

In this chapter, we have studied various aspects concerned with AI ethics, security, and privacy issues. Self-awareness and self-consciousness robots mentioned in Chap. 13 might appear, live, and work among us in the near future. A new set of *AI Code of Ethics* might need not only from human's perspective but also from robotics point of view.

Form a group of 4–5 students and choose to play human or sentient robots:

1 List out 10 major *AI code of ethics* that you think are the most important concern.
2 For each of the AI code of ethics discuss potential risks and problems if is violated.
3 For each of the AI code of ethics discuss how to enforce the regulation (Fig. 14.7).

14.12 Conclusion

In this chapter, we explore one of the most important disputes—*AI Ethics*. It began with three AI ethics golden rules by Prof. Issac Asimov in 1942, to the latest AI HLEG (2019)—the AI ethic guidelines proposed by the European

Commission in 2019. AI has come across numerous changes and revolutions precisely ranging from simple robotic arms in car manufacturing to the latest *robotic companion* and *household robots* in *smart house*. With the advancement of AI, big data and data mining technology and the integration of IoT and 5G technology, self-driving vehicles and autonomous robotics are no more fantasy topics in science fiction, they will exist and live among us in near future. A new set of *code of ethics* for this new age technology is urgently needed than ever before.

AI ethics are mainly focused on the code of rules to protect us and to control robots (or AI systems) from human perspectives in old days. With the latest development of *self-awareness* and *self-consciousness robotics* and AI systems, AI ethics are not only to protect us, but also to protect sentient and self-aware robots' basic rights. All these topics and issues are uncharted territory in civilization. In other words, we all need to prepare and adapt for changes with a new world order. That is, how to live with other species that are equally intelligent or even more intelligent than us. That is a matter we need to consider and well prepared.

References

AI HLEG. (2019). *High-level expert group on artificial intelligence: Ethics guidelines for trustworthy AI*. European Commission. Retrieved May 25, 2020, from https://ec.europa.eu/digital-single-market/en/high-level-expert-group-artificial-intelligence.

Allen, C., et al. (2005). Artificial morality: Top-down, bottom-up, and hybrid approaches. *Ethics and Information Technology, 7*(3), 149–155. https://doi.org/10.1007/s10676-006-0004-4.

Anderson, S. L. (2008). Asimov's "three laws of robotics" and machine metaethics. *AI and Society, 22*(4), 477–493. https://doi.org/10.1007/s00146-007-0094-5.

Asimov, I. (1942) Runaround: A short story. In *Astounding science fiction, March 1942. Reprinted in "I, Robot"*, New York: Gnome Press.

Bennett, C. J., & Raab, C. (2006). *The governance of privacy: Policy instruments in global perspective* (2nd ed.). Cambridge, MA: MIT Press.

Brożek, B., & Janik, B. (2019). Can artificial intelligences be moral agents? *New Ideas in Psychology*. https://doi.org/10.1016/j.newideapsych.2018.12.002.

BS6110. (2016). *Robots and robotic devices. Guide to the ethical design and application of robots and robotic systems* (British Standard). British Standards Institution.

Cervantes, J., et al. (2020). Artificial moral agents: A survey of the current status. *Science and Engineering Ethics, 26*(2), 501–532. https://doi.org/10.1007/s11948-019-00151-x.

Cristianini, N. (2020). *Shortcuts to artificial intelligence*. Retrieved May 25, 2020, from https://philpapers.org/archive/CRISTA-3.pdf.

Ferguson, A. G. (2017). *The rise of big data policing: Surveillance, race, and the future of law enforcement*. New York: NYU Press.

Friedman, B., et al. (1992). Human agency and responsible computing: Implications for computer system design. *Journal of Systems and Software, 17*(1), 7–14.

Goos, M., et al. (2009). Job Polarization in Europe. *American Economic Review, 99*(2), 58–63.

Harari, Y. N. (2016). *Homo Deus: A brief history of tomorrow*. New York: Harper.

Haskel, J., & Westlake, S. (2017). *Capitalism without capital: The Rise of the intangible economy*. Princeton, NJ: Princeton University Press.

Hew, P. C. (2014). Artificial moral agents are infeasible with foreseeable technologies. *Ethics and Information Technology, 16,* 197–206.

Kuflik, A. (1999). Computers in control: Rational transfer of authority or irresponsible abdication of autonomy? *Ethics and Information Technology, 1*(3), 173–184.

Kurzweil, R. (2007). *The singularity is near*. Penguin Books.

Lin, P., et al. (2011). *Robot ethics: The ethical and social implications of robotics* (Intelligent Robotics and Autonomous Agents series). The MIT Press.

Loebner Prize competition regulation. (2003). Retrieved May 24, 2020, from https://loebner03.hamill.co.uk/docs/LPC%20Official%20Rules%20v2.0.pdf.

Macnish, K. (2017). *The ethics of surveillance: An introduction*. London: Routledge.

Marcus, G. (2018). *Deep learning: A critical appraisal*. Cornell University. Retrieved May 25, 2020, from arXiv:1801.00631.

Matthias, A. (2004). The responsibility gap: Ascribing responsibility for the actions of learning automata. *Ethics and Information Technology, 6*(3), 175–183.

McGee, G. (2007). *A robot code of ethics*. The Scientist Magazine. Retrieved May 24, 2020, from https://www.the-scientist.com/column/a-robot-code-of-ethics-46522.

Moor, J. H. (2006). The nature, importance, and difficulty of machine ethics. *IEEE Intelligent Systems, 21*(4), 18–21. https://doi.org/10.1109/MIS.2006.80.

Müller, V. C. (2020). Ethics of artificial intelligence and robotics. In Edward N. Zalta (Ed.), *The Stanford encyclopedia of philosophy* (Summer 2020 Edition). Retrieved May 24, 2020, from https://plato.stanford.edu/archives/sum2020/entries/ethics-ai/.

Rawls, J. (1971). *A theory of justice*. Cambridge, MA: Belknap Press.

Tuchong. (2020a). *AI ethics*. Retrieved May 25, 2020, from https://stock.tuchong.com/image?imageId=920218441399271454.

Tuchong. (2020b). *Robot rights*. Retrieved May 25, 2020, from https://stock.tuchong.com/image?imageId=428507134898077959.

Tuchong. (2020c). *AI surveillance*. Retrieved May 25, 2020, from https://stock.tuchong.com/image?imageId=920062267798061066.

Tuchong. (2020d). *Automation and employment*. Retrieved May 25, 2020, from https://stock.tuchong.com/image?imageId=920766917312905308.

Tuchong. (2020e). *AI and predictive analytics.* Retrieved May 25, 2020, from https://stock.tuchong.com/image?imageId=939676979710263317.

Tuchong. (2020f). *Autonomous smart car in smart city.* Retrieved May 25, 2020, from https://stock.tuchong.com/image?imageId=423875424984629306.

Tuchong. (2020g). *AI codes of ethics.* Retrieved May 25, 2020, from https://stock.tuchong.com/image?imageId=932202095934046211.

Turner, J. (2018). *Robot rules: Regulating artificial intelligence.* Palgrave Macmillan.

Veruggio, G., & Fiorella, O. (2008). Roboethics: Social and ethical implications of robotics. In *Springer handbook of robotics* (pp. 1499–1524). Berlin, Heidelberg: Springer.

Xafis, V., & Labude, M. K. (2019). Openness in big data and data repositories: The application of an ethics framework for big data in health and research. *Asian Bioethics Review, 11*(3), 255–273.

15

What's Next?

As Irving Good realized in 1965, machines with superhuman intelligence could repeatedly improve their design even further, triggering what Vernor Vinge called a singularity.
Prof. Stephen Hawking (Physicist, 1942–2018)

Abstract As a closing chapter of the book, this chapter explores several AI cutting-edge related technologies: singularity and superintelligence, quantum computing, and 6G technologies. First, we begin with singularity in AI and superintelligence concepts and major concerns. Next, we present quantum computing basic concepts and how such innovative theory are true to AI technology. Then, we explore technology beyond 5G—the 6th Generation Communication Technology, its major features, and how these three powerful technologies can be integrated to form a new AI and smart city era.

The *singularity* of AI is the turning point where intelligent machines (or robots) surpass human intelligence and thus become *self-sufficient*. This is also the intersection point where *machine intelligence* replaces human and replicates itself without human intervention.

Is it possible? or just a science fiction or imagination.

We will explore several cutting-edge AI related technologies that will form our future in this last chapter. They are *singularity and superintelligence, quantum computing and 6G technologies*. Although these technologies are

R. S. T. Lee, *Artificial Intelligence in Daily Life*,
https://doi.org/10.1007/978-981-15-7695-9_15

completely independent, the entire AI horizon and the smart city will be changed entirely when they integrate. We will first introduce concepts and major concerns of *singularity* in AI and superintelligence. Then, we will study the basic concepts of *quantum computing* and how such innovative theory is related to AI technology. After that, we will explore technology beyond 5G— the *6th generation communication technology (6G)*, its features, and how these three powerful technologies can be integrated to reshape the new AI era and smart city. Lastly, we conclude with closing remarks—*The future is our choice*.

15.1   Singularity and Superintelligence

The ultimate objective is considered *AGI (Artificial General Intelligence)* implementation since the beginning. That is the development of a *general-purpose system* that can imitate human act and behavior totally with ultimate *self-sufficient, self-conscious,* and *self-awareness* achievement.

The idea of *singularity* (Müller 2020; Shanahan 2015; Kurzweil 2005) is that if the AI's intelligence level attains to a certain point that it has the ability to generate its own intelligence and other intelligent robots without human assistance and intervention—*superintelligence (SI)*. Beyond that moment, SI systems can evolve quickly, improve themselves, and generate more *SI robots (systems)*. After that, anything can happen, and AI development will be out of our control and prediction as what we saw in *Terminator movies series (1984–2019)* and *Transcendence* in 2014 (Fig. 15.1).

Fig. 15.1   Robot with superintelligence (Tuchong 2020a)

The fear that robots will conquer our world finally had awakened human imagination years before computers invention. The optimistic argument from acceleration to *singularity* is proposed by inventor and futurist Mr. Raymond Kurzweil (2005) in his book *The Singularity is Near* who essentially pointed out that according to *Moore's law* that CPU capacity will be doubled in every 2 years. Computers will reach human computing capacity by 2010, that *human mind* can be uploaded to AI systems (robots) by 2030 and *singularity* will occur by 2045. Kurzweil also pointed out at that time all countries and IT companies will be focused on AI development to generate superintelligence AI systems.

Many AI scientists, in fact, estimated that from 2020, onwards actual computing power available to train a particular AI system would double in every 3.4 months, resulting in a 300,000-fold increase, not a sevenfold increase every two years as predicted by Moore's Law.

A general version of this argument (Müller 2020; Shanahan 2015; Chalmers 2010; Bostrom and Yudkowsky 2014) talked about AI systems intelligence improvement (instead of CPU speed), but the key point of *singularity* is still that AI systems take over to further develop and accelerate beyond the human level. In addition to computing power improvement, there are other ways to *superintelligence* such as human brain complete simulation on computers, biological paths or networks and organizations (Kurzweil 2012; Shanahan 2015; Bostrom and Yudkowsky 2014).

Concerns about SI include AI systems (robots) development with *self-awareness* and *self-consciousness*, the transcending of our mind and consciousness into the machine to achieve *immortality* (Geraci 2008; O'Connell 2017). These questions constitute philosophical, epistemological, ethics and moral issues mentioned in previous chapters.

15.2 Quantum Computing

We are entering a historical moment of computing technology—*quantum computing*. IBM's System-Q system was officially launched at CES2019 marking the birth of *quantum computing* in January 2019 (Lee 2020; Moran 2019; Silva 2018).

The contrast with the traditional computers is that instead of using 0 and 1s bits as basic binary representation in data storage and program calculations, quantum computers use *qubits*—two-state quantum superpositions which have greater flexibility in data representation and programming than traditional binary systems (Bernhardt 2019; Hirvensalo 2010; Scherer 2019; Zygelman 2018).

Technically speaking, *quantum computers* will be able to perform calculations in the order of much larger than traditional computers. According to physicist Prof. David Deutsch, quantum computing is the best solution tailored for the modeling of *quantum systems* to tackle complex quantum phenomena, AI for example. The origin of *quantum computing* (Lee 2020), can be traced back to the lecture by theoretical physicist Prof. Richard Feynman (1918–1988) in 1959. He talked about the use of *quantum effect* ideas to make miniaturization of more powerful computers. However, before the calculation of quantum effects, *quantum computing* is still a pure concept in the scientific community. Until Prof. David Deutsch put forward a new concept of *quantum logic gates* in his thesis *Quantum Theory* as a means of using the quantum field in computers, his thesis pointed out that any physical theory can model processes through physical *quantum computers* (so-called *hard quantum computing*) in 1985 (Deutsch 1985; Steane 1998).

Due to the complexity of technology, the path to *hard quantum computing* is not easy. Nearly a decade later, computer scientist Prof. Peter Shor designed a quantum computing algorithm in 1994, which used only 6 qubits to perform basic factorization operations. A few quantum computers have been manufactured since then. The first one was a quantum computer with 2 qubits. It lost coherence after a few nanoseconds in 1998, and then performed negligible calculations. The quantum computing team included IBM and AT&T successfully built a 4-qubit and 7-qubit quantum computer in 2000.

IBM demonstrated its IBM Q-System One at CES 2019, claiming that it was the world's first integrated quantum computing system designed for scientific and commercial use. IBM's 20-qubit system was regarded as one of the leaders in *quantum computing* (Lee 2020; Silva 2018; Moran 2019).

Some people may ask: If we can use traditional computers to model quantum theory and related phenomena, why do we need to build quantum computers?

The answer is simple. If we can use traditional computers with binary representations and representations to model quantum theory effectively, then quantum theory and quantum computers integration will become more direct and effective with the basic structure of quantum computers will fully conform to quantum computers. In other words, if AI can be fully integrated with quantum computing not only calculation speed will be faster, but system architecture of the entire AI system such as the robot system will also be more *realistic* and *closer to reality*. The combination of 0 and 1 includes the realization of emotional robots and *robot consciousness* in future-cognitive

Fig. 15.2 Cognitive computing (Tuchong 2020b)

computing with the help of quantum computing technology to simulate human thinking process (Fig. 15.2).

15.3 6G Technology

What is 6G?

What is 6G? It's a question everyone has been asking since several Asian telecommunications providers announced the commencement of research and development in the past few months. It might seem surreal that attention has already turned to technology at least a decade away with 5G networks just rolled out commercially at the end of 2019.

6G is shorthand for the *sixth generation of wireless networks*, the successor to 5G cellular technology (Nawaz et al. 2020; Kato et al. 2020; Saad et al. 2019; Huang et al. 2019). Most estimates place the widescale 6G deployment no sooner than 2030, though even that date is up for debate. Like all previous generations, 6G is expected to build on 5G capabilities augmented by digital technology advances. That will mean speed and data capacity expansion plus innovations regarding IoT that 5G is widely expected to make practical.

Main Features of 6G Technology

(A) Spectacular Speed and Almost Non-existent Latency

Latency refers to the time between a cause and an effect in the simplest sense. It governs things like the time between hitting to send a text message and the recipient receiving it. Each wireless generation focuses on using different parts of the spectrum (range of radio frequencies) to reduce latency and offer more connectivity, speed, and neat ways for user's smartphone for entertainment. The average latency with 4G stands at around 50 ms. It is about the speed of a lightning flash, but it's still not fast enough to accommodate ultra-sophisticated smart environments or devices. The average latency with 5G drops to one millisecond. It is about five times faster than the flap of a hummingbird's wings. Such a low latency is why the *Internet of Things* may soon become a regular part of our world. 6G is expected to offer speed up to 1,000 times faster than anything 5G can provide. To meet this expectation, it will need to take advantage of radio frequencies that are not cluttered with consumer devices already.

(B) Extreme Connectivity and Sensing

By leveraging the massive bands found at upper gigahertz bounds and lower terahertz frequencies will require more than just faster devices. We will have genius and faster devices. Most speculations expect the sheer speed and data capacity of 6G to transform connectivity on a whole allowing users to connect more devices with smarter and more precise data at a rate that far exceeds anything 5G will offer. US FCC in its report on terahertz waves applications noted that this extreme connectivity would allow for centimeter-level positioning (compared to feet-level accuracy with 4G) and much more sophisticated data centers. This has direct impacts on objects like drone fleets, robotics controls and monitoring devices.

(C) Massive Advancements for Artificial Intelligence

FCC is not the only one seeing the potentials for robotics and smart devices. Some experts claim that 6G's speed and positioning precision will allow AI to process and transmit data at the speed of a human brain finally. Although 5G will make AI more accessible and capable, 6G is likely to transform AI with *disruptive applications* that make its presence in our world much more prominent (Letaief et al. 2019). Nawaz et al. (2020) produced detail research on potential applications for quantum machine learning integration and 6G technology with a comprehensive illustration shown in Fig. 15.3 (Nawaz et al. 2020).

(D) *Comprehensive, Reliable Coverage*

Japanese telecommunications provider *NTT Docomo* published a stunning whitepaper in January 2020 outlining its plan to develop and deploy 6G by 2030. It expected 6G, among many benchmarks to achieve seven-nine reliability (99.99999% reliability). That means continuous coverage

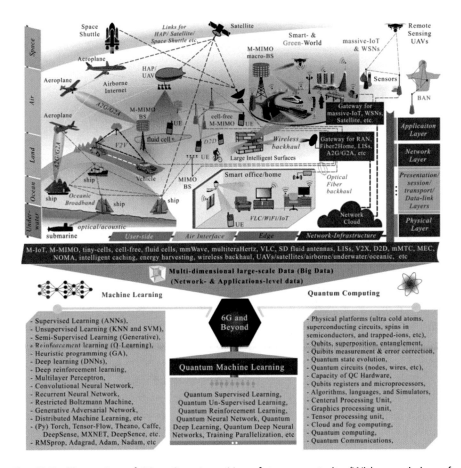

Fig. 15.3 Illustration of 6G + Quantum AI on future smart city (With permission of reproduction of figure from IEEE Access of Nawaz et al. 2020)

anywhere and everywhere effectively. It also described extreme coverage of new boundaries over land, undersea, and even in space.

(E) *Unparalleled Energy Efficiency*

The conversation around 6G is unique that it includes ideas about energy efficiency, something unconsidered by previous wireless network generations. One study from IEEE speculated that 6G will make energy harvesting technologies possible within the network infrastructure. Such efforts will help AI with its massive energy needs sustainable on the network. NTT Docomo takes this idea a step further in its whitepaper. It expects that the next generation of wireless technology will make battery-powered electronics obsolete. Power supply technology using radio signals will lead to more energy-efficient devices with ultra-low power consumption.

15.4   Closing Remarks—The Future is Our Choice

In this book, we had studied one of the most important topics in human civilization—Artificial Intelligence (AI) and how such technology can be applied to our daily activities. AI is not a single technology. It covers different kinds of related computer technologies: machine learning, computer vision, data mining, natural language processing, and agent ontology. Technically speaking, IT or computer technology is only the implementation or how these AI technologies can be applied to different applications. The truth is: AI today should more focus on its core concept—philosophical, psychological, epistemological, and ontological aspects and how these concepts can be interpreted by the latest AI technologies.

Part IV of this book discussed several major concerns regarding AI development: AI and self-consciousness; AI ethics, singularity, and superintelligence. One might ask: If there are many concerns and issues on further AI development, do we really need to do them? Do we really need AI to be so intelligent? Can we just simply control it for further development?

These are in fact philosophical or moral questions than technical or scientific ones. The questions are: *Can we control AI to become truly intelligence? Who have the rights to decide? How can we control human from doing this?* As AI scientist, the author believes we all have the right and duty to investigate the *unknown* and the *truth* of the universe. The future is uncertain and always changes. It is our *free will and choice* to drive our destiny, not the technology.

References

Bernhardt, C. (2019). *Quantum computing for everyone*. MIT Press.

Bostrom, N., & Yudkowsky, E. (2014). The ethics of artificial intelligence. In K. Frankish & W. M. Ramsey (Eds.), *The Cambridge handbook of artificial intelligence* (pp. 316–334). Cambridge: Cambridge University Press.

Chalmers, D. J. (2010). The singularity: A philosophical analysis. *Journal of Consciousness Studies, 17*(9–10), 7–65.

Deutsch, D. (1985). Quantum theory as a universal physical theory. *International Journal of Theoretical Physics, 24*(1), 1–41.

Geraci, R. M. (2008). Apocalyptic AI: Religion and the promise of artificial intelligence. *Journal of the American Academy of Religion, 76*(1), 138–166.

Hirvensalo, M. (2010). *Quantum computing*. Springer.

Huang, T., et al. (2019). A survey on green 6G network: Architecture and technologies. *IEEE Access, 7,* 175758–175768. https://doi.org/10.1109/ACCESS.2019.2957648

Kato, N., et al. (2020). Ten challenges in advancing machine learning technologies toward 6G. *IEEE Wireless Communications* 1–8. https://doi.org/10.1109/MWC.001.1900476.

Kurzweil, R. (2005). *The singularity is near: When humans transcend biology*. London: Viking.

Kurzweil, R. (2012). *How to create a mind: The secret of human thought revealed*. New York: Viking.

Letaief, K. B., et al. (2019). The roadmap to 6G: AI empowered wireless networks. *IEEE Communications Magazine, 57*(8), 84–90. https://doi.org/10.1109/MCOM.2019.1900271

Lee, R. S. T. (2020). *Quantum finance: Intelligent forecast and trading systems*. Springer.

Moran, C. C. (2019). *Mastering quantum computing with IBM QX: Explore the world of quantum computing using the quantum composer and Qiskit*. Packt Publishing.

Müller, V. C. (2020). Ethics of artificial intelligence and robotics. In E. N. Zalta (Ed.), *The stanford encyclopedia of philosophy* (Summer 2020 Edition). Retrieved June 2, 2020, from https://plato.stanford.edu/archives/sum2020/entries/ethics-ai/.

Nawaz, F., et al. (2020). A review of vision and challenges of 6G technology. *International Journal of Advanced Computer Science and Applications, 11*(2). https://doi.org/10.14569/IJACSA.2020.0110281.

O'Connell, M. (2017). *To be a machine: Adventures among cyborgs, utopians, hackers, and the futurists solving the modest problem of death*. London: Granta.

Saad, W., et al. (2019). A vision of 6G wireless systems: Applications, trends, technologies, and open research problems. *IEEE Network* 1–9. https://doi.org/10.1109/MNET.001.1900287.

Silva, V. (2018). *Practical quantum computing for developers: Programming quantum rigs in the cloud using Python, Quantum Assembly Language and IBM QExperience*. APress.

Scherer, W. (2019). *Mathematics of quantum computing: An introduction*. Springer.

Shanahan, M. (2015). *The technological singularity*. Cambridge, MA: MIT Press.

Steane, A. (1998). Quantum computing. *Reports on Progress in Physics, 61*(2), 117–173.

Tuchong. (2020a). *Robot with superintelligence*. Retrieved June 2, 2020, from https://stock.tuchong.com/image?imageId=920874592137904263.

Tuchong. (2020b). *Cognitive computing*. Retrieved June 2, 2020, from https://stock.tuchong.com/image?imageId=947968517154406418.

Zygelman, B. (2018). *A first introduction to quantum computing and information*. Springer.

Printed in the United States
by Baker & Taylor Publisher Services